THE
Great British
SOLDIER

THE
Great British
SOLDIER
A LIVING HISTORY

PHILIP WARNER

David & Charles

Frontispiece: A Grenadier Guard at
            Windsor Castle, Berkshire.

British Library Cataloguing in Publication Data
Warner, Philip
  The great British soldier: A living history.
  I. Title
  355.00920941

ISBN 0-7153-9795-8

Book designed by Michael Head
Typeset by Ace Filmsetting Ltd, Frome, Somerset
and printed in Hong Kong by Wing King Tong Co Ltd
for David & Charles plc
Brunel House   Newton Abbot   Devon

CONTENTS

Introduction   7

1 The Origins of the Modern Soldier   13

2 A Standing Army   26

3 The Redcoats   46

4 A Worldwide War and its Aftermath   63

5 The Napoleonic Wars   71

6 Distant Battles   89

7 The Crimean War   98

8 African and Other Wars   109

9 World War I   125

10 World War II   147

11 Wars since 1945   172

Gazetteer of Museums   179

Index   191

INTRODUCTION

In Britain we are extremely lucky to have a long and varied military history and a large number of museums with displays which make our military traditions vivid and comprehensible. The volume of exhibits in the museums is enormous. They range from uniforms and weapons of the distant past to the tanks, guns, helicopters and military sophistications of the present day. Many museums contain lifelike figures which help the visitor visualise what it was like to have been a soldier at Blenheim, Waterloo, Balaclava, Passchendaele, Alamein or in Korea. Soldiers were not a species apart; they were English, Scots, Welsh and Irish, drawn from all sections of society, and therefore in looking at the British army's evolution through the centuries we are witnessing our own social evolution.

(Opposite) A trooper of the Blues and Royals at Horse Guards, Whitehall.

The Queen's Bodyguard of the Yeomen of the Guard at St James's Palace, London.

This is not history as it is usually taught – a matter of governments, economies, foreign policy and famous names – but the story of life much further down the scale. In recent years the army, like the other services, has been able to be extremely selective, but in the past, many soldiers were press-ganged into the services (the press-gangs operated for the army as well as the navy), were recruited when they were drunk, or they enlisted because the alternative was starvation. The miracle was that however unpromising these men may have seemed at first, army methods and traditions welded the most unsuitable recruits into the finest soldiers in the world.

In the past, soldiers would have carried the identity of the districts from which their regiments were mainly drawn, more than they do today. Consequently, although they became one army, it included a multitude of different components. The Scottish Highland Regiments, for example, were different from the Lowlanders, although, of course, each thought they were better than the others. The Westcountry regiments had no doubts about their inherent superiority over the Easterners, and vice versa. Underpinning this regional pride were the yeomanry regiments with their essentially local traditions. Regiments did not begrudge a word of praise for the fighting achievements of others, but if you asked a soldier – whether he was English, Scot, Welsh or Irish, northerner, southerner, easterner or westerner – where the best regiments came from, the answer soon became predictable.

The military museums of the British Isles bring out the soldier's pride in his regiment, although at the same time they emphasise that the army was totally united. They do not gloss over the hardships, the setbacks, the disease or the deplorable treatment soldiers have often received from mean and ungrateful governments. The military has often been regarded with hostility by the British public and referred to as 'the brutal and licentious soldiery', but the words of Dr Johnson, spoken in 1778, remain as true as ever: 'Every man thinks meanly of himself for not having been a soldier, or not having been at sea.'

Inevitably, most of the exhibits in museums are of the great wars of the present century and many museums contain samples of weapons that have been used by other countries against our own.

An observant visitor will note that although weapons change, the methods they apply vary very little. The ultimate sophistications today are rockets and chemical weapons; both have, however, been used throughout history. For example, rockets were used in medieval times, mainly to carry incendiary materials, chemical weapons have an even longer history, and Greek fire was used in pre-Christian times.

Throughout history, methods for making weapons have been discovered, used, neglected and then forgotten; years, sometimes centuries later they would be revived, or were even rediscovered. An example is Greek fire, a substance whose composition is still unknown, although there are many records to testify that it was an incendiary material which could be projected as missiles or blown through tubes, which would burn into stone or on water, and which seemed unquenchable. Whatever its formula, Greek fire must have used fairly simple and readily available ingredients that were combined, no doubt, in a carefully guarded recipe. Perhaps it contained phosphorus and, as it was extensively used in the Middle East, a form of petroleum. Greek fire seems to have been as deadly a weapon as modern phosphorus grenades and napalm.

The bow has a history which is probably equally long, for 3,000-year-old longbows have been dug up in Swedish peat-bogs. In the Middle Ages the bow was a deadly war-winning weapon, with an effective range not far short of modern rifles. Having the advantage that it was virtually silent, it was used by Commandos in World War II.

Another recent revival is body armour, employed mainly as protection against the sniper's bullet. Body armour of one sort or another has been used from early times, but owing to its weight was usually confined to covering the most vulnerable parts such as the head, groin, chest and shins. The problem with body armour of old was precisely the same as armour on the modern tank. Does one enclose the tank in armour which is thick enough to protect it, but because of its weight slows the tank down and makes it an easier target, or should armour be kept to a minimum, allowing the tank to be speedy and manoeuvrable? Armour of any kind is expensive. A thousand years ago the cost of metal armour put it out of the reach of the ordinary soldier so he wore a jerkin of boiled leather instead. It was not as good a protection as metal, but it would turn a sword cut, or perhaps even a spear thrust, and it left its wearer mobile enough to avoid many blows.

A compromise between weight, flexibility and protection was found by wearing chain mail, but that was still heavy and not a full protection. Consequently, the richer members of the medieval armies were soon wearing plate armour. This was extremely expensive because it required special steel and craftsmanship, but it embodied the very valuable principle of deflecting blows by the angle of the surface. This principle is also used in armoured fighting vehicles today. One of the most successful examples of the use of deflecting surfaces is the Russian T34 tank, which gave the German army an unpleasant surprise when it appeared in 1941.

Army Tank Museum, Bovington, Dorset. A T34 tank can be seen here.

One of the most interesting examples of a skill bridging the centuries is the jointing of body armour. In the Middle Ages this became a highly skilled art that enabled the wearer to protect his joints without losing mobility. In the present century, when American astronauts were planning to visit the moon and walk around in their heavy space suits, scientists in the USA made a careful study of the jointing of the body armour in the Tower of London. They found it very helpful.

The Wallace Collection, London. This collection has magnificent examples of armour.

The horrors of war usually produce great advances in medical science which then has huge resources and, sadly, many bodies on

Royal Signals motorcycles.

Royal Army Medical Corps Museum, Aldershot.
Examples of medical work and instruments, with particular reference to wounds received during campaigns.

which to practise. Medical museums show how many epidemics were controlled, water- and insect-borne diseases understood, and survival techniques developed as a result of war. Surgery before the development of anaesthetics was painful and appalling, but a lesson which was learnt on the battlefield was that the faster an operation can be performed the less is the effect of shock. Human fortitude when people were horribly wounded in the past has lessons that even now we are scarcely beginning to understand. How could badly wounded men lie out on frost-covered battlefields all night (often stripped naked by battlefield scavengers), and yet survive? How could Lord Somerset at the Battle of Waterloo have a wounded arm amputated (without anaesthetic) and then, watching it be carried through the door, say: 'Hey, bring that back a minute. It has a ring on the finger which I need to take off.'

In past centuries disease killed more soldiers than fell in the actual fighting. This gave rise to the saying that until World War I, the doctor killed more than the butcher. However, surgeons often assisted miraculous recoveries by badly wounded men and learnt to apply certain principles of medicine even if they did not always understand why they were effective. Some army doctors were years ahead of their time in the methods they applied. A simple method of purifying water was discovered by an army doctor during World War I. Even more important, perhaps, were the principles of camp hygiene, of which there are many examples. Armies in the field, particularly in the tropics, waged a constant war against disease-carrying insects, and the siting of camps was, whenever possible, on well-drained soil. Then, as now, cook-houses and latrines were situated well away from the sleeping and training accommodation; the cooks had to make sure that food was not contaminated by smoke, and grease, whose presence was inevitable, was prevented from fouling the site by ingenious drainage traps. Nowadays, camp-sites have to be concealed and camouflaged and cannot be sited in the most hygienic places, but fortunately modern antiseptics can do much to counteract the dangers of otherwise unhygienic positions.

Military history is therefore not only a chronicle of battles against other nations but also includes the battle against disease and all threats to health. The blood transfusion service, which has saved the lives of many unfortunate victims of accidents, was largely developed by armies. Protection against dangerous fumes, as essentially required for firemen, was learnt in the appalling conditions of gas warfare during World War I. In some museums you may see examples of primitive gas masks.

Virtually all forms of the modern rescue services have been developed by armies (or navies or air forces). Lost ramblers, trapped climbers and rash or inexperienced sailors have often owed their lives to the skills that were developed by armies to save

A 'cutaway' Centurion tank at Bovington.

A unique motorcycle (1926) in the
Museum of Army Transport.

the lives of their own men when they were engaged in hazardous operations.

One of the most interesting aspects of military life has been to test men and women to the limit and in the process to learn that human endurance is vastly greater than had been imagined. On sale in many museums are books of survival; the lessons they teach have nearly all been learned from soldiering. Among the techniques they disclose are many which make camping holidays much safer and more enjoyable; this is not merely a matter of how to pitch tents or light fires but how to remain warm, dry and reasonably happy when the climate is at its worst.

When visitors look at guns and tanks in museums, they may not at first realise that the techniques which produced modern weapons have also provided such valuable assets as modern steel and many alloys which are used in today's domestic life. Some of these discoveries, which have been of enormous benefit to civilians, were made when experimenters were looking for an alternative metal – for example, stainless steel was first discovered when an alloy, thrown onto the scrap heap, was found not to have rusted.

Having discussed a few of the interesting features of modern military museums, we must now go back to the beginning and look at the evolution of the modern soldier and his weapons.

CHAPTER 1

THE ORIGINS OF THE MODERN SOLDIER

THE REGIONAL REGIMENTS

In this book the regiments and corps which make up the British army will be referred to many times. Each regiment has a local association which preserves its traditions and records and stimulates recruiting. Those characteristics and traditions are of enormous importance. Although the corps, being newer than the regiments, have less strong regional ties because they draw their membership from all over the British Isles, they too have established links with the area around their depots and training grounds. This happens because many of their members buy houses and settle in the district where much of their military life has been spent. The corps also go to great lengths to establish friendly relations with the local community by 'open days', sporting contests, dances and assistance with local community projects. Thus, Arborfield and the surrounding district contains many former members of REME, Blandford and much of Dorset is the home of numerous former members of the Royal Signals, the RAOC has members around Blackdown, and many Royal Engineers live in Kent because of their long link with Chatham.

Because regiments were originally raised from certain areas, usually counties, they tend to embody, and preserve, the traditional characteristics of those regions. Thus the Middlesex (now part of the Queen's) have become noted for their quick Cockney humour. 🔫 But the fact that they tend to be light-hearted has not made them any less formidable as a fighting force, as their nickname 'The Diehards' testifies. Devon people may seem mild and inoffensive, softly spoken and slow to take offence. However, when a whole division of the German army, desperate to reach Paris in 1918, encountered a single battalion of the Devons on the way, thus outnumbering it by fifteen to one, they found nothing mild or inoffensive about the reception they got – they were firmly stopped in their tracks. 🔫

Army humour could fill a book of its own, and every regiment and corps has stories about various mishaps which have a lighter side to them. The Welsh and the Scots are seldom short of an answer. Sometimes a witty remark can defuse a potentially dangerous situation; at others it exposes the pomposity of the questioner.

A few years ago the refuse collectors in Glasgow went on strike and, within a short time, there were piles of rubbish in the streets and the hazard of disease. The Royal Scots were given the unpleasant task of clearing the mess and while doing so were filmed for television. The soldiers had a horrible task. Bags burst when they were picked up, rats scuttled around and the smell was

🔫 **Middlesex Regimental Museum, London N17.** Holds the regimental relics, weapons and uniforms.

🔫 **Devonshire Regimental Museum, Exeter.** Holds the regimental relics, weapons and uniforms.

Museum of Queen's Own Highlanders, Fort George, Inverness-shire. A display of weaponry and other mementoes, held in an eighteenth-century fort.

The Black Watch on ceremonial parade at Edinburgh Castle.

nauseating. A reporter asked one of the soldiers condescendingly, 'I don't suppose you like this job very much.' The Scot looked at him. 'Well, at least it's stopped me biting my nails,' he replied cheerfully.

Scottish regiments have an awesome reputation on the battlefield. Germans who experienced their bayonet charges over No Man's Land during World War I nicknamed the kilted Highlanders 'The Ladies from Hell'.

Over the years various regiments have earned a reputation for prowess with certain weapons. Most cavalry regiments became experts with lances, which are difficult weapons to handle successfully; it took two years to train a lancer. Guards regiments have a proud tradition of being resolute under pressure, and their highly

Changing of the Guard at Buckingham Palace.

disciplined regiments – the Coldstream, the Scots, the Irish and the Welsh – have sometimes fought until the whole unit has been virtually wiped out. However, the Guards are not merely first-class infantry. During World War II the War Office decided to form a Guards Armoured Division.  When the order was issued to begin this project, there were many doubters. The Guards, it was agreed, were first-class infantry but it was felt that they would never master the mysteries of mechanical warfare. Furthermore, they were such large men, owing to the physical standards required of recruits, that they would never be able to fit into the cramped space that is available in tanks. The critics, however, were wrong, for not only did the Guards adapt themselves to tanks, but they became one of the best armoured units of the war.

Obviously, all attempts to link regional backgrounds to regimen-

Guards Museum, London SW1. Here you can see a collection of exhibits of all the Guards regiments, held in an underground museum.

tal performance must be very much a matter of generalisation. But although generalisations tend to be scorned by the academic world, they usually contain a large nugget of truth. It is sometimes said that in the past some regiments have become so tribally inward-looking that this has been damaging to efficiency; however, during the major wars of this century and in the subsequent regimental mergers, the 'tribes' have admitted and absorbed members of other 'tribes' without losing any of their identity and fierce pride in it. In fact, the newcomers were often even more 'tribal' than the original members. However, when several regiments merge the early stages may be problematic. For example, during World War II a Midlands regiment which had been virtually wiped out was made up to its original strength by two large drafts, one from Scotland and the other from the Westcountry. Inevitably, the soldiers of each draft spoke with a heavy regional accent. For the first two or three weeks the situation verged on the chaotic. Undoubtedly there were occasions when the Westcountry soldier genuinely could not understand the orders of the Scottish NCO and vice versa, and the Midlanders could understand little of either. Gradually matters eased: phraseology became more precise and orders were enunciated rather than grunted from the back of the throat. But a hard core of the regiment who did not care for obeying commands they disliked, took advantage of the linguistic difficulties. When they were reprimanded for not having completed a task, they would advance the plausible excuse: 'I couldn't understand what the sergeant [or corporal] was saying.' The problem looked insuperable, but it proved not to be. At the end of a month the commanding officer issued a decree: 'All orders will be obeyed whether they are understood or not.' From then on, they were.

The reason why people from different parts of the British Isles are different in manner, speech, appearance and military characteristics lies deep in history. To understand and appreciate these regional differences, we need to take a quick glance backwards, even as far back as thousands of years. We are all the product of our environment and ancestry and the more we know about those two influences, the better we understand and appreciate ourselves and others.

Knowing about ourselves, the county we live in and the relics around us adds greatly to the pleasure and interest we experience in everyday life. Ancient roads, earthworks, castles, local customs and characteristics all have a history that still influences us. To get the best out of our holidays, leisure and recreation, we need to know and understand something of this background.

Crickley Hill, Cheltenham, and Almondsbury, Huddersfield. Earthworks dating back to about 3,000 BC can be seen at both these sites.

FROM PREHISTORY TO THE CELTS

The earthworks which mark the British landscape indicate that warfare has existed in the British Isles from about 3,000 BC. Before that date the population was so small that men's fighting ability was directed against the wild animals which threatened their existence rather than against fellow humans. Over the centuries, nomadic peoples from the Continent came to Britain in search of new hunting-grounds or more fertile areas for their primitive agriculture. Most of these nomads arrived on the south and east coasts, and drifted across the country, settling here and there. Although over two thousand years has passed since these early immigrants arrived in Britain, in certain rural areas there are still people whose features and other characteristics, such as the colour of their hair and eyes, indicate a distant tribal origin.

In the period after 1,600BC, the first half of which was known as the Bronze Age and the second the Iron Age, the pace of immigration quickened and brought in some very accomplished fighting tribes. As each wave of immigrants arrived, its predecessors were pushed further to the north and west – to Wales, Ireland and Scotland. Mountainous or marshy regions gave refuge to people who had been driven off their former lands by more numerous and better-armed opponents. We know a surprisingly large amount about the people, habits and weapons of those times; much of it is to be seen in the many archaeological museums around the British Isles (see Christopher Somerville's book in this series, *The Great British Countryside*). When the first settlers, who may not originally have been warlike and had no alternative but to flee or become slaves, went to the Westcountry or to the North, they soon adopted fighting techniques which enabled them to preserve their independence. They became agile as a result of living in rugged country and developed throwing weapons – precursors of the modern rifle and gun – to prevent their opponents from reaching close quarters. They became experts at ambush and would lure their opponents into peat bogs or into valleys where they could be trapped and then crushed with boulders rolled down the hillside. They developed what are now called 'hit-and-run' tactics by which they could goad their enemies into making mistakes; they would wear down their opponents until they were hungry, weary and vulnerable to defeat and slaughter. The fighting methods which were developed by the hill dwellers became the guerrilla-style tactics of the SAS, the Commandos, the Rifle Brigade and many other regiments that are skilled in skirmishing and escape-and-evasion. Many highly prized techniques of modern regiments would have been familiar to what are now roughly grouped together as the 'ancient Britons'.

Many of the famous fighting tribes had women as their chieftains: Boadicea, queen of the Iceni in the first century AD, is famous for the way she defeated the Roman armies in a rebellion which caused the legions surprising military difficulties, but others have been equally formidable. In the nineteenth century, regiments of the British army took pride in the fact that they were fighting for 'the Widow of Windsor', as Queen Victoria was called by them; and many regiments today have colonels-in-chief who are women. When the Queen appears on horseback at the Trooping the Colour, or Princess Anne takes part in field exercises with the Royal Signals or other regiments of which she is colonel-in-chief, their presence can be traced back to a custom which existed a thousand years before modern regiments came into existence. The British army is steeped in traditions, some going so far back into the mists of time that their origins seem almost untraceable. In more recent history there have been many redoubtable warrior queens: Margaret of Anjou, Queen Elizabeth I of England, Catherine the Great, and others have inspired armies to great military deeds. Some female war-leaders were not queens, however: Joan of Arc, for example, is a folk-heroine in France – and whether the British forces in the Falkland Islands would have been so successful without the driving resolution of the then Prime Minister Margaret Thatcher is questionable.

Another tradition of the British army – and, of course, other armies – is the alliance between religion and military prowess. The Celts were tall, fair-haired, belligerent invaders who settled in Britain before the Romans came but who were eventually driven into northern Wales and Scotland. (The southern Welsh, who tend to be small and dark, came from Iberian stock.) The Celts introduced religion to the battlefield. Their priests, who included both men and women, inspired the warriors to superhuman feats of endurance and courage. When Celtic armies confronted their enemies, the Celtic priests danced in front of them and cursed their opponents. This was the beginning of psychological warfare, now known as 'psyops', by which the army tries to intimidate the enemy by propaganda and other methods of undermining their resolution. The Celts also introduced the custom of covering their swords, shields and helmets with intricate decoration, like the engraving on a modern shotgun; the custom was also continued on field artillery pieces until recent times. The Celts, too, were exponents of chariot warfare. These men (and women) were expert charioteers, who could bewilder opposing armies by sudden forays and swift retreats; in World War II this tactic was used by jeeps and armoured cars. Of course, the Celts had their setbacks: their chariots were trapped in cunningly concealed enemy trenches, and priests and monks were slaughtered on the battlefields where they had become overconfident of their own invulnerability.

Intelligence Corps Museum in Ashford, Kent. Here you can see exhibits which show the history of intelligence gathering.

THE ROMANS

A different system of military management arrived with the Romans and many of the Roman methods are still in existence. The Romans were the first people to grasp the importance of good military roads. Whenever possible their roads were straight, even if it meant building viaducts and avoiding hills. Roman chariots bequeathed us the standard gauge which is used on the modern railway. We know this because of the Roman habit of cutting two grooves in the edge of their roads to allow the chariots to reach the raised surface of the round more easily – these are found to be

142cm apart. However, the Romans left the British much more than a gauge, as you may see if you visit any museum that specialises in Roman heritage. ☞ They built viaducts, some of which were used for canals, they constructed bridges, some of which are still in use, and they even dug tunnels. The Romans attached so much importance to military road-building that the first-class roads they constructed could encircle the earth twice; if all the minor roads they built were included, Roman roads could go around the earth ten times. The Fosse Way (Lincoln to Exeter) and Watling Street (Kent to North Wales) remain, in parts, major trunk roads.

Roman road construction was a masterpiece of engineering. At the base were large stones set in mortar; above this was a layer of smaller stones and above that was a layer of even smaller stones which were set in mortar. The top layer, a mixture of gravel and sand, made an easy marching surface for the legions. Although a light horse-drawn chariot could cover 75 miles (120 km) in a day, the foot soldier was not likely to march more than 20 miles (32 km), except in an emergency. From their successes and failures the Romans learnt many military lessons. One was that it was unwise to march until dusk and then pitch camp, for the soldiers would be tired and might not be able to make a secure defensive position before nightfall. Roman soldiers therefore ended their daily march in the early afternoon. They carried their construction tools with them and on a march would quickly construct a fortification with a deep ditch, a bank made of the soil from the ditch and a row of stakes on top of the bank. Some of the roads and camps the Romans built have lasted for two thousand years, even though they have been neglected for much of that time. Their roads were much easier to march on than the cobbled roads (*pavé*) in France, which British soldiers came to hate during World War I.

Among the many valuable lessons that the British army has learnt from the Romans, not least is the value of hard work and continuous weapon training. Their soldiers were highly disciplined and had to practise with their weapons for most of the day. They became expert at getting to close quarters and destroying their opponents, but they were skilled with the chariot (a skill they had learnt from the Celts) and the javelin. They had bows too, but did not consider them as important as their other weapons.

Although they were familiar with the sling, the Romans did not specialise in the use of the hand-sling, preferring to recruit auxiliaries from countries in the Middle East where expert slingers could be found. The Romans themselves had plenty of experience of the power of the sling, for many of the tribes they encountered were extremely accurate with the cheap and deadly weapon. Slingers had been in existence since pre-Christian times (as the story of David and Goliath bears witness) and we know that they were used in the battles the Romans fought in Britain because archaeologists have found many thousands of sling-stones in the ditches outside some of the great earthworks the Britons tried to defend. ☞

Chesterholm Museum, Hexham. A fascinating museum of relics excavated from Vindolanda Fort, which was the base for Roman soldiers stationed along Hadrian's Wall.

Maiden Castle, Dorset. Visitors to this well-preserved earthwork fort will be able to gain a good idea of the desperate battles which took place in the 1st century AD, when no quarter was expected or given. Some of the relics, such as skulls split in half with axes, or bones with cross-bow bolts still stuck in them, leave little to the imagination. Another earthwork worth visiting is Spettisbury.

Roman relics and graphic displays may be seen at the Vindolanda Museum, Northumberland.

Chesterholm Museum, and Clayton Memorial Museum and Chester Roman Fort, Northumberland. These are all fine examples of Roman wall forts along Hadrian's Wall.

Although they used bows and arrows, the Romans preferred to put the same principles (of propulsion by tension) to use in the construction of larger weapons, which were the predecessors of the modern howitzers and field guns. Giant catapults, which were able to project large and heavy stones over several hundred yards, were used to batter a breach in the walls of solidly built fortresses. These weapons, for which the 'elastic' was either human hair or horsehair, were extremely accurate. When they were not required to demolish castle walls, they were used in much the same way as a modern howitzer or trench mortar to send a projectile at an angle of 45°, thus clearing outer defences and carrying some lethal cargo, perhaps incendiary material, on to the unfortunate occupants.

Although the Romans built fortified towns, they did not expect to be besieged in them. Consequently, they had wide gates through which their soldiers could emerge rapidly to deal with strife. Their towns possessed straight, uncluttered interior lines which facilitated the mustering and despatch of troops. Once a country had been occupied and pacified, their troops reverted to what is now known as a 'fire brigade' role (troops rush to crisis points), although nowadays they are likely to be airborne in such contingencies.

However, although the Romans believed in using their troops as mobile strike-forces, they also had a strong belief in the value of walls for keeping out unwanted visitors. One of the best examples is Hadrian's Wall (Wallsend on Tyne to Solway Firth), which was built to keep the more intractable Scottish tribes from periodically raiding the settled and prosperous lands of northern England. It is possible to walk along much of Hadrian's Wall and also to visit some of the old wall forts. By the time the wall was built the Romans were using large numbers of mercenary troops from countries they had conquered. These soldiers could look forward to a pension, some land and Roman citizenship if they served their time satisfactorily; but such promises did not stop them inscribing the walls of their forts with messages which indicated their homesickness; when he was half-frozen and bored with peering through the Scottish mists and rains, wondering if a resourceful enemy might be about to attack, the mercenary might well have regretted the decision which had taken him from his homeland.

When the Roman empire was in decline there was an increasing tendency to rely on fixed defences. Roman armies built walls not merely around towns or across a section of the countryside (for example, Hadrian's Wall), but around whole countries. The history of warfare suggests that such defences are rarely effective: the Great Wall of China was a stupendous exercise but a military failure; in the present century other 'impregnable' fortifications have been breached: the Maginot Line in France, the Siegfried Line in Germany and the Todt fortifications of western France are among them.

With a huge empire to control, with warlike tribes on their frontiers, and with restive people within, it is not surprising that the

Romans made the best use of the only communications that were available. Messages were despatched by runners or horsemen, but a quicker method of giving warnings was by smoke signals. Thus signal stations were established at strategic points of the Roman empire; at Scarborough the signal station is believed to have been built to give warning of the approach of seaborne raiders.

THE VIKINGS

Although the inhabitants of Britain had bitterly resented the arrival of the Roman soldiers in Britain in the 1st century AD, they were unhappy to see them leave some 450 years later. Under Roman protection the British had neglected the martial arts and at first they made mistakes which invariably proved fatal. Some of the raiders who now descended in large numbers on the shores of Britain had made smaller incursions long before the legions left this country and in consequence the Romans had set up defence lines against them in what became known as 'The Forts of the Saxon Shore'. These ranged from Brancaster in Norfolk to Portchester in Hampshire. The larger ones, such as Portchester, are so solid and powerful, with their bastion towers and their mountings for great catapults, that it has been suggested that they were originally built by a rebellious Roman governor who had decided to assert his independence from Rome and expected a full-scale Roman fleet to be sent against him – the forts seem altogether too powerful for the ships used by the Jutes, the Angles and the Saxons. But powerful or not, they gave no protection to the Britons when the Roman soldiers were no longer there to garrison them.

A few Britons made the fatal mistake of trying to defend the forts and were soon massacred. Experience taught them that the only way to survive was to become guerrillas and to take refuge in the mountains or swampy areas of the British Isles; there they settled, alongside the descendants of their predecessors who had been driven from their own homelands. It should be remembered that, although the Romans had made enormous progress in improving the roads and towns of the British Isles, vast areas were still almost uninhabitable. Rivers were full of tangled vegetation and easily overflowed into the surrounding areas. Much of what is now East Anglia was marsh, which afforded useful refuge points. Similarly, there were large swampy areas in the West Midlands and the North.

But soon it was the turn of the Angles, the Saxons and the Jutes to come under threat. The new arrivals, who came from Denmark, Norway and Sweden, were known as the Vikings. They seemed to have a fiendish delight in battle and a ferocious cruelty to accompany it. At first they were content with murderous raids, which began in 787, in which they plundered the towns and abbeys that the Anglo-Saxons had built in Britain. Their technique was the swift, sharp commando raid, which was later given the title 'butcher and bolt'. They were experts at mobile warfare. They would arrive

Pevensey Castle, Sussex. (on the A259). The scene of one of these incidents. It also contains a Norman Castle and some World War II modifications.

without warning, sail their boats up whatever rivers or creeks they could find, then capture any horses which were available and use them to raid far inland. Then they were gone, leaving behind them bodies, burning homes and a swathe of destruction. However, after taking part in a few raids, some Vikings decided that Britain was a better environment than their homelands which were bare and overpopulated in relation to their resources, and they settled into a new way of life. When later invaders came to plunder Britain the new Viking settlers made common cause with the Anglo-Saxons into whose community they were by then being absorbed.

In order to defend themselves against invaders, the Viking immigrants joined the Anglo-Saxons in making a form of defence called a 'burh'. This was a fortified enclosure, which was situated usually in a cultivated area rather than on a hilltop. As cultivated areas rarely have natural defensive advantages, the burhs had to be formidable constructions. Usually they consisted of an enclosure which was surrounded by a ditch and thorn palisades. They were large enough to be a refuge when an attack threatened and were too strong for a Viking raiding party to destroy. Gradually these temporary burhs became larger and formed permanent settlements. Usually they were near the coast, as at Scarborough or Aldeburgh, but many were also built far inland as defence against the raiders who became full-scale invaders and penetrated deeply into the country. When the Vikings settled, their own town names ended in 'by': Grimsby and Whitby are obvious examples on the coast, but the Vikings penetrated far into the Midlands too, as the name Rugby indicates.

We can easily identify the position of many of them today by town names: thus, Edinburgh was the burgh in which Edwin was the chief.

From this period of battling to survive derive many of the military characteristics of the British soldier. He learnt the value of mobility, the importance of signalling communication, the means by which an enemy army can be defeated by a series of strongpoints and, above all, the need to be constantly alert against a surprise attack. This experience would be invaluable when, many centuries later, he was required to fight in countries far distant from his own – for example, India, Canada and Africa.

WILLIAM THE CONQUEROR

At the end of the first millennium, Anglo-Saxon Britain seemed to have absorbed its military immigrants and had become the Anglo-Saxon Kingdom of England. However, a new and irreversible change was coming from across the Channel. Although England was still subject to raids from the warlike Scandinavians, those old adversaries fought with methods and weapons which were understood and could be countered. The invaders who arrived on 14 October 1066 were of a different kind. They came from a wide area but mostly from France and were then known as the Normans. They fought on horseback, instead of using their horses merely for transport. They had short bows and carried long shields and long swords. They wore iron helmets and armour which consisted of

Battle (7 miles/11km north-west of Hastings). The critical battle between the Normans and the Saxons, where Battle Abbey now stands, was fought here.

Beaumaris Castle, Anglesey – the ultimate in medieval concentric defence. Begun in 1295, it took 28 years to build.

chain mail. Chain mail was made of hundreds of circular links which could turn a sword or spear thrust but also allowed the wearer a degree of mobility. Chain mail was not proof against the crushing blow of a Saxon battle-axe, but it was very effective against the cruder arms of the Saxon rank and file.

The Norman army which defeated the Saxons at Hastings in 1066 also brought a concept of warfare which eventually led to the British regimental system. Although he had won an important battle at Hastings, William the Conqueror knew that winning one battle did not mean that he had won a war. He therefore set about subduing the rest of the country and then holding it down with a well-organised system of military control. That system was to allot parts of the newly conquered country to his most loyal and successful supporters and then to demand a form of rental from them in periods of military service. He encouraged them to build strongpoints – known as motte and bailey (mound and enclosure) castles – and to enrol their local subjects into military forces which would be available when required. This system, which established a personal relationship between the local baron and his people, foreshadowed the later link between the regiment and its recruitment area. The standard period of military service, which every baron was required to give to the monarch, was forty days. To this day, territorial forces go into the field for a set period of military training every year (although for less than forty days).

There was, however, a considerable drawback to this forty-day period of service because tenants were often called up just when they were needed for either sowing or harvesting. Furthermore, when the forty days had been completed, the armies were entitled to disperse whether a campaign had been completed or not. Gradually, there crept in a modification of the system by which the barons could pay an indemnity ('scutage') to the monarch for not providing soldiers; the monarch would then use this money to hire mercenaries.

Mercenaries were more satisfactory than the feudal levies provided by the barons: they were available throughout the year and could therefore take part in campaigns overseas (such as in France), they were better disciplined because soldiering was their career and they were better trained because they had plenty of time to practise with their weapons. It might have been thought that because mercenaries were not tied to their leaders by anything stronger than pay and the loot they might obtain from a successful campaign, they would be liable to desert in crises. In fact, the reverse was the case. Mercenaries would often fight for their paymaster literally to the last man. No doubt they were often offered inducements to join the opposite side, but there seem to be few occasions when they did so.

OF FOREIGN ORIGINS

It is worth remembering that in past centuries 'British' armies were often a mixture of French, Spanish or other European troops. Similarly, British soldiers often served in the armies of other countries. Some of the Anglo-Saxons who were defeated by William the Conqueror at the Battle of Hastings in 1066 escaped and enlisted in the élite guard in Byzantium (modern Istanbul).

Even the word 'soldier' has a foreign origin. It comes from the old French word *saude*, meaning 'pay'. A 'saudier' was originally a man who fought for pay; he was therefore different from a man who bore arms because of his feudal obligations.

Many other foreign words have been introduced by soldiers and become Anglicised. When soldiers fought in North America they adopted the words 'wigwam', 'moccasin' and 'squaw'; when they were in Malaya they acquired 'sarong', 'amok' and 'gong'; from the Turks they took 'yoghurt' and 'kiosk'; from Persia they took 'caravan' and 'shawl', and from their three hundred years in India they acquired a host of words, some of which, like 'bungalow', became entrenched in the English language and some, like 'jildi' (hurry), have gradually disappeared during the forty years since British soldiers left that country.

CHAPTER 2

A STANDING ARMY

THE NEW MODEL ARMY

Soldiers who stand for hours on parade waiting for VIPs to appear could well believe that the words 'standing army' mean that soldiers spend most of their time on their feet. (The old sweats have a saying, 'Never stand when you can sit, and never sit when you can lie down.') But, of course, a standing army is one which is never fully disbanded. The first standing army was created by Cromwell in 1645 and was called the 'New Model'. Its numbers, 7,600 cavalry and 14,400 infantry, make it seem small by modern standards, but it was a very professional force none the less. It had cannons and handguns and was properly trained. Many of its soldiers were battle-experienced. It was properly fed, clothed and paid.

The New Model Army was therefore a very different force from any armies that had preceded it, which, although often successful, had usually consisted of a motley array of troops. Before Cromwell's New Model Army, soldiers wore clothing which could not be described as uniform, although the archers usually had leather jerkins, and caps into which they put their bow-strings when it rained. Their food supplies were erratic and their staple diet was often salt

The bloodiest battle ever fought on English soil took place at Towton, near Tadcaster, Yorkshire, between the Yorkists and Lancastrians on 27 March 1461 in a snowstorm. It lasted from dawn until dusk and caused 28,000 casualties.

(Above) The Battle of Bosworth, Leicestershire (1485) marked the end of feudal warfare and established the Tudor dynasty.

Poitiers, 19 September 1356 – the surrender of the French King (Jean) to Edward III. Elaborate courtesy, but a very bloody battle.

27

The Battle of Pinkie (near Edinburgh) 1547 was between the English and the Scots. One of a long series of battles of great ferocity, this one was won by England.

Castle Museum, York. The life of the Civil War soldier can be explored in detail here.

herrings which were transported in barrels and were far from tasty. The soldiers were liable to go for long periods without pay and then have a chance of plunder in which they could reward themselves with food, wine and trinkets, all of which soon disappeared. The systems of command and communications were erratic. When British armies had won some of their greatest victories, as at Crecy (1346), Poitiers (1356) and Agincourt (1415), they had been almost at the end of their resources.

Cromwell knew exactly how an army should be organised. The New Model Army had been compulsorily enrolled by order of parliament, in contrast to the volunteer system in the king's army. After the Civil War had been won by parliament in 1646, the New Model reverted to voluntary recruiting and by 1651 it had expanded to 70,000 (over three times its original strength). It had a proper command structure in which separate units had to obey *all* the orders from above, not just the ones they were prepared to accept. In its ranks was a balance of artillery, engineers, infantry, catering and supply services. The cavalry tended to be better educated and wealthier than the remainder, but in all branches the officers, many of them from the lowest ranks, were men who had been promoted on merit. As they had learnt their trade in campaigning and battle, every one had proved himself capable of discharging the duties of the rank he held.

However, the most unusual and novel feature of the New Model Army was that it was motivated by religious zeal. All ranks read the bible and listened carefully to sermons; hymns and psalms were sung regularly and fervently, and all the troops believed that God was on their side. They would not have been amused if they had been told that the pagan Druids had held a similar belief. Furthermore, the New Model was a highly disciplined army; punishment was severe for many misdemeanours which would not be considered an offence at all in today's army. Card-playing, gambling, swearing or intemperance would get a man into serious trouble, even though Cromwell's own youth was not as blameless as it was thought to be.

From the late 1640s the musketeers carried either matchlocks or flintlocks which were about 4ft (1.2m) long. The musketeers wore red coats, breeches and stockings, and felt hats. Their bullets were carried in the bandolier that was slung across their chests and the powder for firing them was in the powder horn carried on the shoulder. The musketeer's belongings were carried in a knapsack which was worn on his back. Musketeers were steady men who would stand fast in their ranks even when they were being harried by cannon fire, and they would not waste ammunition or opportunities by firing prematurely at long range.

Near the musketeers would be pikemen who wore helmets, breast- and back-plates. They carried a sword and a 16ft (5m) long pike, one end of which would be braced in the ground when a cavalry charge was expected. A squadron of pikemen could stop a cavalry charge, but they found it difficult to defend themselves if the cavalry stopped short, wheeled and then discharged their pistols (called dragons because they spouted fire) into the faces of the waiting pikemen. The cavalrymen who fought this way became known as dragoons; they wore body armour, carried a sword and a pistol, and were expert at skirmishing and reconnaissance.

An early wheel-lock pistol (sixteenth century). The wheel was wound up by a winding spanner. When the dog (hammer) was released, a piece of iron pyrite was pressed against the wheel as it revolved; this ignited the powder charge and fired the gun.

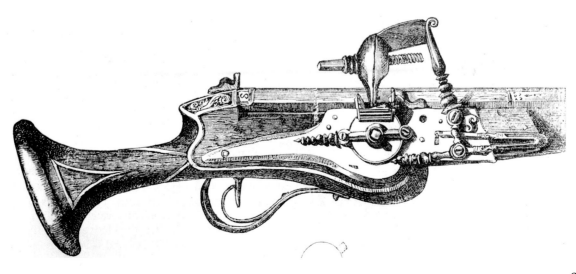

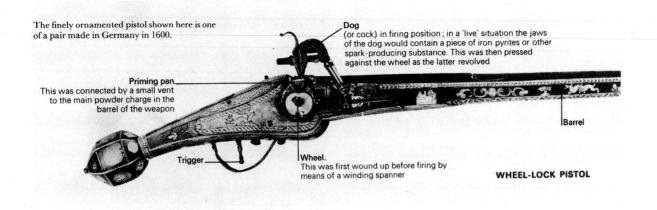

The finely ornamented pistol shown here is one of a pair made in Germany in 1600.

Dog
(or cock) in firing position ; in a 'live' situation the jaws of the dog would contain a piece of iron pyrites or other spark-producing substance. This was then pressed against the wheel as the latter revolved

Priming pan.
This was connected by a small vent to the main powder charge in the barrel of the weapon

Barrel

Trigger

Wheel.
This was first wound up before firing by means of a winding spanner

WHEEL-LOCK PISTOL

A 1600 wheel-lock pistol.

The heavy cavalry also wore body armour and carried pistols and swords, but their horses were larger and heavier and their swords were longer. The artillery, which came in a variety of sizes, was organised with teams of horses and oxen. The artillery carried round-shot and barrels of powder, but their main purpose was to drag the heavy guns into positions from which they could direct their shot at castle walls or into the ranks of the opposing army. When the artillery encountered the enemy it was often difficult to distinguish friend from foe, for both sides tended to dress and look alike. To ensure, therefore, that they were not attacking their own side, the artillery would wear a distinguishing token that had been especially chosen for the occasion – for example, an oak leaf worn in the cap. Medieval armies, which were cased from head to foot in armour, encountered the same problem, which gave rise to the practice of painting their shields with the coats-of-arms that have become the distinguishing marks of families, regiments, schools, colleges and towns.

When an enemy army was sighted, the New Model Army would prepare for action in a manner which would seem to a modern soldier to be lengthy and unnerving. The musketeers would take up station between the pikemen, who would thereby give them protection against a sudden cavalry charge. The musketeers were arranged in three ranks: the front rank was kneeling, the second was bending forward and the third was upright. Initially, musketry drill was not well organised for once the muskets had been discharged there was a time lapse before they were all reloaded, during which the enemy cavalry would be advancing towards them. Later, an elaborate drill movement was devised to enable one rank to fire, then step aside and let another fire while the first rank reloaded. Eventually, the drill was perfected so that as soon as one rank had fired another was ready to step forward and take its place. At one time this manoeuvre required four ranks, as reloading was a lengthy business. In the present century army organisation has reverted to three ranks for parade and marching purposes. Infantry no longer goes into battle in carefully organised ranks, but is dispersed to offer the smallest target to enemy machine-gun, artillery or aerial attack, but the column of threes still has many uses.

Military routine has varied little during the centuries that armies have fought battles. Out in front have been the skirmishers, looking for the enemy and reporting back on his probable strength and disposition. Then the artillery has opened fire at long range. When the opposing armies have closed up, the heavy cavalry has protected the flanks and watched out for an unexpected attack from the rear; meanwhile the infantry has fought a dogged battle in the centre. If the cavalry saw a chance to put in a quick and decisive charge on enemy infantry which was already beginning to give way, it would do so, and if the enemy began to give way all along the line the

This exhibit in York Castle Museum gives a good impression of pikes and halberds which were used for several centuries in Europe.

cavalry would be ready to turn defeat into a rout. If, however, it was the enemy who seemed to be getting the upper hand, the cavalry would try to break up the attack and escort its own infantry to safety. The other parts of the army had tasks appropriate to their skills, such as building or destroying bridges and fortifications, tending the wounded, organising supplies of food or ammunition where they were needed, and relaying information to the commanders.

No battle is ever won until it is completely over, as many an over-confident commander has discovered to his cost. During the first English Civil War of 1642 to 1645, the Royalists often lost valuable gains by being over-impetuous. At Edgehill, Prince Rupert pushed back the opposing wing of the Parliamentarians and then pursued them for miles; when he returned he discovered that the battle had not been won by the Royalists and his absence from the field had put them at a serious disadvantage. Having failed to learn his lesson, he made the same mistake at Naseby in 1645, which was an even more critical battle. Once troops scent victory they are not easy to control, but there are innumerable examples throughout history of headlong charges losing their cohesion and bringing defeat to themselves and their own side.

THE GREAT CIVIL WAR

Although civil wars, in which a nation tears itself apart by internal fighting, are considered the worst type of war, eventually they tend to produce a more unified country and a more efficient military system. As few people nowadays are familiar with the course of the Great Civil War, and therefore wonder why it occurred and what happened on its battlefields, this momentous event is described below.

The Civil War began in 1642 because the king, Charles I, refused to accept the fact that, as a constitutional monarch, he had to abide by the decisions of parliament which had gradually been growing in strength and authority. Instead, he believed in the 'divine right' of kings which gave him the authority to rule as he wished. Having tried, and failed, to arrest the leading members of parliament, Charles abandoned London and rallied an army at Nottingham. With this show of strength he hoped to overawe the Parliamentarians and then to rule the country without opposition. Parliament, however, did not intend to be over-ruled in this way and organised its own military forces.

The armies first met at Edgehill (Warwickshire) on 23 October 1642 in a drawn battle. Charles could, however, have marched to London immediately afterwards and re-established himself, but he was too dilatory and by the time that he reached the outskirts of the city (at Turnham Green) parliament had organised a sound defence; he therefore fell back to Reading to collect a stronger force.

Various battles between the rival supporters of the king and parliament then took place in different parts of the country. The Parlia-

Edgehill, near Kineton, Warwickshire. Using a map, you can trace the course of the Battle of Edgehill.

Naseby, Northamptonshire. Here you can 'walk' the battlefield and imagine the battle held there between the Royalists and Parliamentarians.

mentarians (nicknamed Roundheads from their close-cropped hair which made a sharp contrast to the flowing locks of the Royalists) captured most of the towns along the east coast, but in the west, the north and Wales the Royalists gained the upper hand. Charles set up his own parliament in Oxford, using the college hall at Christ Church for the purpose, and also established a mint in order to make coins out of the college and family plate, which was melted down. Many of the battlefields of the Civil War – Edgehill (Warwickshire), Chalgrove (Oxfordshire) [🔫], Lansdown (near Bath), Roundway Down (Wiltshire) and Marston Moor (Yorkshire) – remain much the same as they were then and tourists and holiday-makers can plan enjoyable excursions to visit them and imagine the long-past battles. They may also occasionally see events which leave nothing to the imagination. Some years ago Brigadier Peter Young, a distinguished Commando who was also a first-class military historian, hit upon the idea of creating a Royalist (and later Parliamentarian) army to re-enact the battles of the Civil War. His army, 'The Sealed Knot' as it was called after a Royalist emblem, was immensely popular and enrolled whole families, including women and children, into its ranks. Dressed up in (mainly home-made) Civil War uniforms, equipped with cavalry and volunteer stuntmen, with proper artillery firing noisy blank ammunition, the Sealed Knot was soon a popular draw at festivals of all kinds. It was filmed and televised and soon had branches throughout the British Isles. It continues to thrive. [🔫]

The Battle of Naseby, which took place on 14 June 1645, was the decisive battle which gave Parliament's forces victory in this, the first phase of the war. It is said that Naseby was not so much a battle which Cromwell won as a battle which Prince Rupert, the Royalist cavalry commander, lost.

After Naseby, Charles I was forced to surrender but he refused parliament's terms for his reinstatement. Instead he intrigued with foreign powers to restore him and eventually the war began again in 1648. This phase was soon over and resulted in the beheading of Charles and the execution of many of his supporters in the following year. Charles had undoubtedly been misguided and obstinate, but his parliamentary opponents were equally fanatical. The only good to come out of the Civil War was that the principle of constitutional rule was firmly established and the army's authority was laid down. There were also numerous bad effects, such as the loss of priceless silver and gold plate, the vandalism of churches during which parliamentary fanatics broke the stained glass because it was 'Papist', and the damaging of many castles which had held out against Parliament's forces. The policy of damaging castles was called 'slighting' and its traces may be seen clearly today: at Bridgnorth and Caerphilly [🔫] castles are towers which, although they were blown up, refused to fall and lean at an alarmingly unsafe angle to this day, three hundred years later. Medieval cement seems to have been more efficient than the modern variety.

Oxford. Earthwork defences are still visible to the north and east of Oxford.

Castle Howard, York. One of the major battles of the Civil War, Marston Moor (2 July 1644), is now re-enacted at this lovely stately home, where there are better facilities for spectators. Newbury, in Berkshire, where two battles took place, is another favourite venue.

Bridgnorth, Shropshire, and Caerphilly, 9 miles/15km north of Cardiff. Here you can see two castles with leaning towers.

THE RESTORATION

The New Model Army won many more victories, notably in Scotland and Ireland, where there was considerable opposition to Cromwell's parliamentary forces. It was soon respected and feared outside Britain. In 1655, Cromwell sent it to assist Louis XIV of France in his war against Spain. Louis XIV would later be engaged in almost continuous war *against* Britain, but at this stage he looked a useful ally against Spain, whom Cromwell hated because its Catholicism and ambitions were as strong as his own Protestantism. The New Model Army captured Dunkirk from the Spaniards in the Netherlands (which later would be called Belgium and Holland) and for a while Dunkirk became a British possession.

By 1660 Cromwell was dead, Britain had become an incompetent military dictatorship and there was much disagreement among the military leaders. There was still Royalist opposition, which was now beginning to raise its voice, and there was a widespread feeling that the present state of affairs could not and should not continue. The most able of the military leaders, General George Monk, who commanded the army in Scotland, marched south, entered London and summoned the surviving members of parliament which had last met in 1642. A new parliament then came into being and decided to restore the monarchy, provided that the new king, Charles II, would agree to rule constitutionally. One of the new monarch's first decisions was to maintain on a permanent footing the four regiments of foot and three of horse which had formed the backbone of the New Model Army. Among them was General Monk's Regiment of Foot, also known as 'the Coldstreamers', after Coldstream, the little Berwickshire village in which they had originally been formed. In 1661 they became the Lord General's Regiment of Foot and in 1617 the Coldstream Regiment of Footguards. In 1817 they were designated their present title, 'The Coldstream Guards'.

However, it should be remembered that this was not yet a proper regular army – that would not begin until 1689. These early regiments were referred to as 'Guards and Garrisons'. The king was their nominal head but they were commanded by a commander-in-chief or captain-general. Parliament was very wary of establishing a force that could be used by the king as Charles I had tried to use his troops, and was equally determined not to have a repetition of the last stages of Cromwell's government, when he had divided the country into twelve districts and put a major-general in charge of each of them.

However, seven regiments were not enough to fulfil all the commitments Britain had at home and overseas and some others were added, eventually bringing the total to eight of foot and four of horse (16,000 in all).

Charles' wife, the Portuguese heiress Catherine of Braganza, had brought as her dowry the two trading posts of Bombay and Tangier. Bombay was soon leased to the East India Company for a rent of £10 a year, but Tangier was retained for its useful harbour and naval base. Two regiments, called the Tangier Horse and the Tangier Regiment respectively, were raised to defend it from hostile neighbours. However, Tangier was soon shown to cost more to defend than it was worth and in 1687 it was abandoned. The Tangier Horse thereupon became the Royal Dragoons (now amalgamated with the Royal Horse Guards, 'the Blues', to become the Blues and Royals). Both these regiments had won battle honours when they were in Tangier. Although today Tangier would seem to be a delightful place to be stationed, in the seventeenth century it was notorious for its disease and appalling living conditions.

In the series of amalgamations which took place in 1971, the Queen's absorbed the Buffs, the Royal West Kent Regiment, the Royal Sussex, the Royal Surrey (formerly East Surrey) and the Middlesex, although the last of these was reduced to a token number of half a dozen. ☞ Amalgamations and mergers have never been popular in the army but the 1971 series, which was forced upon the army by reduced commitments and national financial emergencies, produced a storm of indignation. However, most of the merged regiments now seem to have settled into their new roles. Few of the present members of the regiments remember their former titles or different status, but recently there have been some misgivings in higher authority about the suppression of county links, now that recruiting figures are falling. Recruiting is usually at its best when the army has dangerous tasks, such as the Korean War or Malaya Emergency, but in the 1990s is bound to be affected more by the birth rate of the 1970s than by any other factor.

☞ **The Queen's Regimental Museum, Kent, and The Queen's Royal Surrey Regimental Museum, Guildford.** These two fascinating museums hold a host of relics from their regiments' histories.

ARTILLERY

Artillery in various forms had been in existence long before Cromwell's army gave a spur to its further development. From carvings, friezes and writings, we know that a form of artillery existed as long ago as 500BC, but at this period and for the next thousand years it was of the type described earlier – that is, the projection of heavy stones by means of giant catapults. Models of these have been made for recent films and their performance has been impressive. ☞ It is said that a Roman catapult at Dover could project 100lb (45kg) to distances of up to a mile (1.6km). The missile in the catapult was launched from a sling and that sling could be used for even more intimidating cargoes. A form of bacteriological warfare was produced by slinging dead and diseased horses into the interior of a beseiged town or castle. Incendiary material could be sent in the same way and a form of terrorism was employed by launching captured spies or hostages back into their own lines. These practices could not be continued once cannon came into use but are closely akin to modern tactics when aircraft and helicopters can drop napalm, defoliants or even more unpleasant material.

Early artillerymen were well aware of the effect of a barrage and in consequence would use as many as several hundred of their giant catapults on certain occasions, usually sieges of towns or fortresses.

☞ **Aldershot Military Museum.** Here you will find a picture of a giant catapult.

A medieval siege; catapults like this had been in use since Roman times.

Fig. 20

VILLE

CAMP DE L'OST

A. The Cat.   B. The Pulley.   C. The Catapult.
D. The Crossbow-men.   E. The Wooden Tower and Drawbridge.

There is a record of a siege in 1266 at Kenilworth Castle in Warwickshire when the missiles from each side flew through the air in such large numbers that they often collided in flight – a form of anti-ballistic missile defence which, though unintentional, was very effective.

Although gunpowder was known in ancient China, where it was used for fireworks and rockets, it does not seem to have appeared in Europe until the middle of the fourteenth century. A hundred years later it was in regular use. Pioneer artillerymen were very inventive and intelligent. In approximately 1450 there was a weapon called a ribaudequin. It consisted of tubes bound together and could fire in salvos while it was rotated. This principle was used in World War II in the German Nebelwerfer, a weapon with eight barrels which could launch 80lb (36kg) rockets at the rate of one a minute to a range of 6,000yd (5,500m).

Imperial War Museum, London. One of the deadly World War II rockets can be seen here.

The most intractable problem of early artillery was how to construct a barrel that would not burst when the charge was fired. Many did, with disastrous consequences. The men operating these guns came to be called 'gynours' and from that word we have developed the word 'gunner'. It may also be the origin of the word 'engineer'. The gynour's task was to ram a charge of gunpowder down the barrel and then to ignite it through a touch-hole in the breach. However, the occasions when the touch-hole blew out the burning gunpowder, badly wounding or killing the gynours, soon led to a powder train to be substituted.

Artillery soon developed into two specific types: the gun and the bombard. The gun fired directly at the target; the bombard – the predecessor of the howitzer – launched its shot high into the air, thus clearing any obstacles on its way to its destination. Although most of the shot used by both types was solid, there were experiments to make it even more lethal when it arrived at its target. Some cannon balls were filled with gunpowder but, as this needed to be ignited before the missile was launched, it was seldom successful when it was fired from a gun. A more successful experiment was the grenade, a small bomb held in the hand before it was ignited and thrown at the enemy. Men who were skilful at throwing grenades became known as 'grenadiers'; they were usually tall enough to be able to see suitable targets and most regiments had at least one company of men who were specially chosen for this purpose. However, the British Grenadier Guards did not acquire their name from their grenade-throwing skills, but from the fact that they defeated the French Imperial Guard at Waterloo in 1815. Before that they had been 'The First Regiment of Foot Guards' and in 1660 had been 'The King's Royal Regiment of Guards'.

In the early stages of artillery, guns had descriptive names, such as culverin, which was derived from the old French word *couleuvrine*, meaning 'snake'. Guns seemed to resemble snakes, both having long bodies and a mouth which was capable of effecting a fatal injury. Even at a much later date, many guns were carefully decorated to give the muzzle a very realistic, snake-like appearance. Smaller guns were known as *demi-culverins*.

These terms remained in use until it became more convenient to classify the guns by the weight of the shot they could fire – for example, 16- or 25-pounder. Certain guns then became extremely popular with their users because of their accuracy and reliability: the 13-pdr and 18-pdr of World War I are one example. The 25-pdr of World War II was also regarded with affection by those behind it. Occasionally, certain guns were named after their designer or place of origin. Continental countries classified their guns by

Imperial War Museum, London, and The Royal Artillery Museum, Woolwich. These excellent museums both have fine collections of weaponry.

The Artillery Museum at Woolwich.

their calibre (internal diameter): the French 75mm was sometimes used by the British army. After 1949, when Britain became a member of NATO, the standardisation of weapons meant that in future all British guns would be known by their metric calibre, and nowadays we are accustomed to speaking of 120mm or other sizes of gun.

As guns became more powerful, recoil became an increasing problem and it is interesting to visit museums and see the various devices that were used to counter this, or even to use it for automatic reloading and firing.

Royal Artillery Museum, Woolwich. Examples of anti-recoil devices can be seen here.

Although guns were vital to armies, they created their own special problems over transport. To draw them horses were needed, and to maintain the horses in first-class condition good forage was essential. Horses also required farriers to keep them well shod and veterinary services to keep them healthy. We shall look into this more closely when the part played by cavalry in the army is examined.

As soon as guns were in regular use it was clear that the manufacture and transport of ammunition was going to be a serious problem. Behind the guns themselves there was a long tail of transport containing ammunition, forage and replacements. The faster a gun could fire, the heavier was its requirement for ammunition. During World War I, in which artillery barrages preceded any offensive, there was soon a serious shortage of ammunition. It became clear also that heavy guns could only move on properly made roads. Hundreds of years after the Romans had pioneered the system of military road-making, European nations suddenly realised that unless they, too, made proper roads, their guns would soon sink so far into the ground as to be useless. In certain areas of the Western Front during World War I guns disappeared entirely into muddy shell-holes that had been created by the combined effect of earlier bombardments and heavy rain.

JAMES II AND KILLIECRANKIE

Charles II reigned for twenty-five years until 1685, but his brother James II who succeeded him reigned for only three years. James II was determined to re-establish Roman Catholicism as the principal religion of Britain, which meant that Protestantism had to be suppressed. James believed in the principle of the divine right of kings which gave the monarch the authority to rule exactly as he or she pleased and he behaved as if he was unaware of the fact that the holding of these beliefs had led to the Civil War and the execution of his father, Charles I. Eventually, parliament decided that James would have to be deposed by force and replaced by his sister's husband, a Dutchman, Prince William of Orange. James fled from the country without putting up more than a token resistance, but once in exile did as much as possible to create difficulties for his successor on the English throne.

Prince William, who now became William III of England in 1688, already had plenty of trouble on his hands, for Louis XIV of France wished to extend the frontiers of France to the Rhine. To gratify that ambition he needed to annex the Netherlands and certain other territories to which he had no legitimate, logical or historical claim. William III was an extremely capable soldier but he realised he could not defeat Louis on his own. In consequence, he built up a formidable alliance against which Louis battered his head in vain for nearly twenty years. However, before we look into the series of wars which then resulted for British soldiers, we need to examine a battle which had considerable influence on British fighting tactics – the battle at Killiecrankie in Scotland, in 1689.

Among the exiled James II's supporters was a Scottish nobleman, John Graham of Claverhouse, Viscount Dundee, who collected an army of Highlanders and announced his intention to restore James to the throne. In 1689 William organised an army under a general named Mackay to fight the Scots, but in the narrow pass of Killiecrankie, near Dunkeld, ✍ the English army was ambushed. At that time bayonets were fitted into the muzzle of the rifle after the initial volleys had been fired. This took time and while the English soldiers were performing this operation at Killiecrankie, a number of Highlanders, armed only with swords and shields, swept down on Mackay's unfortunate soldiers and cut them to pieces. Dundee was killed in the moment of victory and the Scottish insurrection, lacking a distinguished leader to hold the chieftains

Killiecrankie. Killiecrankie is on the A9, near Blair Atholl. The National Trust for Scotland Centre on the pass has a presentation of the battle.

together, no longer constituted a threat. But the lesson about bayonets had been learnt. Following this battle Mackay produced a rifle with clips on the side to which the bayonet could be fitted. It would therefore no longer be necessary for soldiers to stop firing in order to fit a bayonet; instead they could continue firing until the enemy came into bayonet range. Killiecrankie was the last battle to be fought when a force armed only with swords overcame an army with muskets. The Scots, whose victory at Killiecrankie had led to the introduction of the modern type of bayonet, took readily to the new weapon themselves, and thenceforward the threat of a bayonet charge by a Scottish regiment, accompanied by the sound of pipes and war-cries, would chill the bravest heart.

THE EARL OF MARLBOROUGH

John Churchill, Earl of Marlborough, who was to become one of Britain's greatest generals, was the chief instrument in the defeat of Louis XIV of France. As already mentioned, Louis wished to expand the frontiers of his country to the Rhine, and so tried to annex the Netherlands and certain other territories to which he had no legitimate claim. The result was a period of wars throughout Europe, which was not freed from the threat of French domination until 1713. Churchill was awarded the earldom and later the dukedom of Marlborough in recognition of his skill in holding together the Grand Alliance against Louis XIV and in winning spectacular victories in the field. He is an ancestor of Sir Winston Churchill, architect of victory in World War II.

There is a saying that it is easier to deal with your enemy than with your allies, for at least you know what your enemy is planning to do. Seldom has a general had such a divergent alliance to hold together; it included Austrians, Dutch, Danes and Hanoverians, and many minor princelings who cared less about the general plan than achieving the best for their own domains with the minimum of effort. Marlborough's English troops only numbered about one quarter of the whole and some of the time he would conceal his plans from his army in case any sector should raise an objection.

Marlborough's special skill as a general, which brought him decisive victories at Blenheim (1704) ⟨image⟩, Ramillies (1706), Oudenarde (1708) and Malplaquet (1709), was in out-manoeuvring the opposing forces. Instead of hurling his army headlong into battle, he would outflank and baffle his opponents and then attack with great ferocity at what he knew was their weakest point. For the battle of Blenheim he took mobile artillery, the first general to do so; the most effective of his guns were the 3-pdrs. One of the first principles of war is to try to surprise your enemy and Marlborough generally achieved this aim. Another principle is that of concentration of force to breach the enemy's position. Once that is achieved, the gap is widened by every possible means and the attack must not slacken until victory is clearly won, otherwise the enemy may recover and counter-attack. An enemy who might have been

Blenheim Palace, Woodstock, Oxfordshire. This was one of the Duke of Marlborough's rewards for his various successes in battle. It is a magnificent stately home, now fully open to the public.

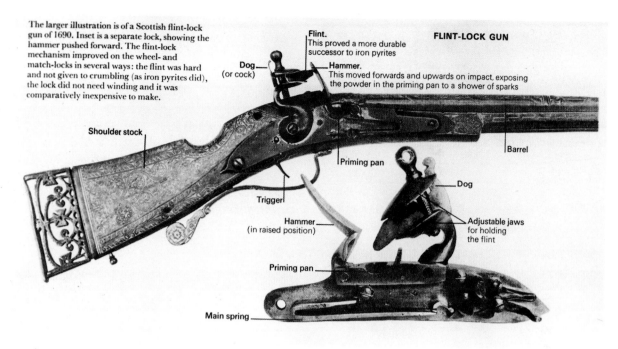

The larger illustration is of a Scottish flint-lock gun of 1690. Inset is a separate lock, showing the hammer pushed forward. The flint-lock mechanism improved on the wheel- and match-locks in several ways: the flint was hard and not given to crumbling (as iron pyrites did), the lock did not need winding and it was comparatively inexpensive to make.

FLINT-LOCK GUN

Flint. This proved a more durable successor to iron pyrites

Dog (or cock)

Hammer. This moved forwards and upwards on impact, exposing the powder in the priming pan to a shower of sparks

Shoulder stock

Barrel

Priming pan

Dog

Trigger

Adjustable jaws for holding the flint

Hammer (in raised position)

Priming pan

Main spring

A flint-lock gun.

expecting Marlborough's army to appear and attack on one side would often find it attacking from the other.

Marlborough also helped to develop the abilities of the British soldier by his encouragement of marksmanship which, until his time, had been haphazard and had wasted a great deal of ammunition. Marlborough insisted that his fusiliers (musketeers) must wait until their targets were close and then make sure that they hit them. He ordered that they should fire in volleys, a well-aimed volley having a devastating effect. He used cavalry to smash opposition by the weight of the charge, rather than taking it to a position where the horseman could use his sword and lance. Marlborough showed himself to his troops on every occasion and was their hero. Although he was a general, the troops felt that he was ready to take the same risks and hardships as themselves. He was their 'Corporal John'; he understood them and they understood him (or they thought they did) and as long as he was in charge they would have victories with the minimum of casualties.

Among Marlborough's victories is one which few people attributed to him later: the capture of Gibraltar. In 1704 he sent a small army with the fleet to capture Toulon. It failed in its mission, but it turned aside and captured Gibraltar, which has been a British possession ever since, despite many attempts by other nations to capture it.

Unfortunately, Marlborough's glittering military career ended in disgrace when he was dismissed from duty for taking bribes from contractors and appropriating government money.

MARLBOROUGH THE TACTICIAN
Marlborough was one of the first generals to appreciate that the

first requirement for a successful campaign is detailed planning. This was particularly evident in his first and best known victory at Blenheim on the Danube, which was referred to earlier. Early in 1704 his army was stationed on the lower Rhine to protect Holland against a French invasion. However, in the meantime Louis XIV had deployed another army in the south, and this was moving forward to a position where it could confront and defeat the Austrians, who were one of Marlborough's allies. Marlborough, who was well aware that the Dutch would not tolerate being left undefended while the British army went to protect the Austrians, had to conceal his real intentions. He therefore had to give both the French and the Dutch the impression that he was manoeuvring to defeat the French in the north, while in fact he was making a forced march to confront the French army in the south. The march took six weeks and he arrived with the 40,000 troops with whom he had set out, without losing a single man or a horse. Dumps of food, horseshoes, forage, boots and other necessities had been established on the way and this careful planning gave his men confidence that victory was assured. Their belief was not mistaken: the enemy were defeated by brilliant tactics and Marlborough's army was able to return to the Netherlands before the French army in the north realised what a splendid opportunity had just been missed.

One of the most remarkable of the soldiers who fought in Marlborough's army was an Irish woman named (originally) Christian Cavenaugh. She had joined the army as a private in 1693 and was wounded in an early battle; however, on that occasion she managed to conceal her true sex. But in a later battle (Ramillies) when she was wounded again, the doctors were not deceived. As she had then served for thirteen years, it was considered that, although she had deceived the army, she had earned a pension of one shilling a day for life. She lived for another thirty-three years and married three times. Her first two husbands were killed in action but her last husband died as a Chelsea Pensioner. Both Christian Cavenaugh and her third husband were buried in the hospital cemetery. There is some doubt, however, whether she deceived her fellow soldiers about her sex or whether they assisted her in concealing it from the authorities. There were other formidable women who travelled with the armies, cooked, washed, nursed and bore them children, but in moments of danger would seize the nearest weapon and give as good an account of themselves as any man.

THE EIGHTEENTH-CENTURY SOLDIER

By this time the standard dress for a soldier was a red coat, woollen breeches, stockings, shoes, shirt and sash. Red had been adopted for the coat because it was the colour worn by most of Henry VII's troops when he had won the Battle of Bosworth Field and thus the throne of England in 1485. It was said to be popular because it was a bright, cheerful colour, which made the soldier look smart and attractive to potential lady admirers; it was therefore a useful

Bosworth, Leicestershire. Here you can see displayed the whole of the battlefield, and an account of how it was fought.

(Above) Royal Hospital, Chelsea.

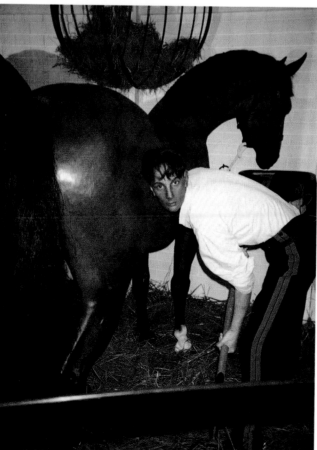

A stable scene from the museum of the Queen's Own Hussars, Lord Leycester Hospital, Warwick. Ingenious use of mirrors gives the impression of a busy stable.

aid to recruitment. On the practical side, it was said that it was adopted because it did not show the blood of a wounded soldier and was therefore good for morale. However, as it was usually accompanied by white breeches, this argument does not seem a very strong one. Red clothing was finally abolished in 1898; by that time it was realised that such a bright colour made a soldier an easily discernible target. However, red continues as an army colour for sporting events and it is also worn on ceremonial occasions, such as public duties.

The essential difference between the army in the eighteenth century and the present day is that all officers' commissions were bought and sold, and a regimental commander was responsible for the clothing, pay and equipment of his men. He was allotted a sum of money by the government for this purpose, but how much of it he used was his own business. Some regimental commanders were rich and took pride in equipping their regiments with their own colours, rather like a racehorse owner today, but others were both greedy and corrupt.

Although the army theoretically consisted of volunteers, there was very little that was voluntary at times of war. Men were released from gaol on condition that they joined the army; in some parts of the country the sheriffs were empowered to seize anyone who had no wife or children or any obvious means of support. This resulted in an army which needed firm handling if it was to be successful and the discipline was severe, although not perhaps any more severe than the punishment code for erring civilians. There were twenty-five offences for which a man could be put to death (robbery being one) and flogging was a regular occurrence. Executions and punishments were compulsorily witnessed by the other members of the regiment to deter them from committing similar offences. In the early period of the regular army, canes were used for flogging, but a century later the notorious cat-o'-nine-tails was employed. Men were tied to a triangular frame and given up to two thousand lashes, and were inspected periodically by the regimental doctor to make sure that they had not died in the process. Young soldiers, frequently officers, fainted as the blood flowed freely down the back of the unfortunate victim, but nothing stopped the punishment except death. Soldiers had been enlisted for life, so if they ran away they were liable to be executed if they were recaptured.

The cat-o'-nine-tails was not introduced until the mid-eighteenth century and was abolished by the mid-nineteenth. It consisted of nine lengths of knotted whipcord attached to a wooden handle. After a dozen or so strokes a man's back would bleed freely and eventually look like raw meat. Over the years regulations were made to limit the number of strokes to two hundred, which was severe enough.

Flogging was not limited to the services: men and women were also publicly flogged, though not with the cat. Vagrants, thieves and

Guards Museum, Wellington Barracks, London. Here you can see a cat-o'-nine-tales.

prostitutes were often victims. In relation to the punishments of medieval times, flogging probably did not seem a particularly severe punishment. In the army it was probably regarded less as a punishment than a lesson in how to behave when wounded and in severe pain and a useful preparation for battle. Soldiers who had never been flogged were despised by their fellows, just as later a man who had never been on a 'charge' (committed a crime) was not thought to be a real soldier. One soldier who had received a flogging, pulled himself upright, saluted his commanding officer and thanked him.

There were, however, a few examples of humanity in the treatment of soldiers. At the end of his period of service – that is, when he was no longer of any use to the army on account of infirmity or disablement, a man would be entitled to a small pension. The average period of service appears to have been about twenty years.

Some old soldiers were selected for residence at the Royal Hospital, Chelsea, which was established by Charles II in 1682. Originally it consisted of ten officers and one hundred disabled soldiers who were given a uniform, a basic diet and a small weekly pay allowance. ☞

The Chelsea Hospital was not, however, the earliest or the only 'hospital' for former soldiers. Over a hundred years earlier the Earl of Leicester had founded the Lord Leycester Hospital in Warwick. It had been established to care for twelve needy, disabled soldiers and continues in that capacity to the present day. ☞

'Hospital' in this sense meant much the same as the modern word 'hostel' or 'almshouse'. Hospitals in the modern sense, where the wounded or seriously ill might be cured, hardly existed. Anyone who had the misfortune to be admitted to a 'hospital' was as likely as not to acquire an additional disease which might be more serious than the cause of his original admission. Up to the time of the Crimean War (1854 –1856), and in some places even later, medical treatment in the army was extremely harsh. This is not to say that all the doctors were brutal and uncaring: there were some who performed magnificently in virtually impossible conditions. It needs to be remembered that up to the nineteenth century all medical treatment was crude and ill-informed. Doctors could set broken limbs and check bleeding, but the causes of most diseases were a mystery to them. A widely practised remedy for a fever was to bleed a patient, as this was supposed to draw off the bad blood from the brain. In certain cases this was effective in that it reduced congestion caused by over-eating, heavy drinking and lack of exercise, but there must have been some literally bled to death by ill-informed surgeons or, worse, amateurs such as barbers or farriers. Astonishingly the public was as deluded as the doctors and submitted willingly. Shock, pregnancy, fevers, or venereal diseases were all considered likely to benefit from this cure-all.

The Royal Hospital, Chelsea, London. The Royal Hospital continues to this day and its members may often be seen on ceremonial occasions, where their scarlet coats, cocked hats and military demeanour make them stand out clearly. Visitors are welcome to visit the hospital.

The Lord Leycester Hospital, Warwick. The Lord Leycester Hospital also contains the regimental museum of the Queen's Own Hussars, which holds many interesting relics.

CHAPTER 3

THE REDCOATS

THE COMMISSIONED ARMY

To this day people still believe that commissions in the services are obtained by purchase, in spite of the fact that newspapers frequently carry advertisements for young men and women to join the navy, army or air force, and to be paid while training for commissions. Of course, no one is likely to obtain a commission unless he or she has a certain degree of education, and as 'education' means staying at school until the pupil is approximately 18 years old, that means that he or she must be supported by their family rather than finding a job and becoming self-supporting. To that extent, obtaining a commission, like obtaining a degree, does require some personal sacrifice, even today.

Until 1939 potential officers were required to pay an annual fee for attendance at the Royal Naval College, Dartmouth, the Royal Military Academy, Woolwich, the Royal Military College, Sandhurst, or the RAF College, Cranwell. The only other way of obtaining a commission was to serve in the ranks for many years and then, having reached a certain level *and* obtained certain educational certificates, to be given a direct commission, probably as a Quartermaster Lieutenant.

The days when commissions could be bought outright ended in 1871 with what were known as the Cardwell Reforms. Edward Cardwell, who was Secretary of State for War, reorganised the whole army between 1868 and 1872 and among his most important reforms was the abolition of the absurd system by which both commissions and promotion could only be obtained by purchase (except when an immediate superior was killed in action); even infants could hold the colonelcy of regiments, if their parents could afford to buy it for them. The same system applied in other countries, too.

The origin of the unjust system of buying commissions lay in the ancient practice by which landowners were expected to provide armed forces for the service of the Crown and also to equip and clothe them; soldiers would recover their expenses from the plunder and ransoms they obtained during the course of a campaign. When the army began to settle into its later form ransoms were no longer given but they acquired plenty of money from loot, fines and indemnity payments. An army campaigning overseas could find rich rewards if the battles were successful, many of the gains coming from the general share-out of booty.

During the early stages of the standing army a cornet or ensign (the equivalent of today's second lieutenant) would have to pay an average of £500 for his commission; the figure varied from the more expensive cavalry regiments to the more economical infantry, but

nevertheless it was the equivalent of several thousand pounds in modern money. The ensign also paid for his own uniform, which was expensive. For every 'step', as it was then called, the officer had to pay a further substantial sum; for the very rich this was not a problem, for they could afford to pay the equivalent of £100,000 for command of a fashionable regiment and a similar sum for new uniforms according to their whims. For officers who liked soldiering as a career and were successful, this total blockage on further promotion without payment was extremely frustrating. Against their enormous 'purchases' outlay, officers would receive a daily rate of pay; £1 would be a customary average for the colonel and 5s (25p) for the ensign. These figures must be multiplied by approximately twenty for a modern equivalent.

An embroidered mitre cap worn by a grenadier officer in a Scottish infantry regiment, seventeenth to eighteenth century.

Eventually, some officers became generals. When they reached that rank they ceased to receive any payment unless they were serving in a war or in some special appointment. However, as a general was almost invariably the colonel of his former regiment (and perhaps some others too) from these he would receive a major portion of the regiment's earnings. Today a regimental colonel is an honorary appointment.

Clearly the difference in the backgrounds of the privileged and rich officer and the men in his regiment was enormous. An officer would never take off his elegant clothes and show his men how certain duties or sports were performed. Instead he would stand aside (if he bothered to attend at all) and watch a non-commissioned officer – usually a sergeant or corporal – perform training or other exercises.

An officer was expensively dressed, possibly scented, and would have been mounted on a fine-looking horse. The only occasions when a soldier was likely to hear an officer speak was when he appeared before him to answer for a crime. Although the Earl of Marlborough, the outstanding eighteenth-century general, talked to his soldiers, he had few imitators because the average officer had nothing to say to his men. In certain regiments to this day men are not encouraged to be talkative. 'Sir,' is the required answer when they receive instruction or advice. Anyone who says, 'Yes, sir' or 'No, sir' is told not to be garrulous.

Nevertheless, when his regiment went into action, the soldier, of whatever period, knew that whatever the circumstances, his officers would lead from the front and be indifferent to personal danger. Surprisingly, when the regiment was suffering hardship or defeat, on occasions like the Peninsular War in the early nineteenth century, the officers often withstood the harsh conditions better than the men. This may have been because officers had the advantage of stronger constitutions, which they had acquired as the result of better and more plentiful food throughout their lives.

Rifleman Harris, a former Dorset shepherd boy, served in the 95th Foot (later the Rifle Brigade and now part of the Royal Green Jackets). After being discharged from the army in 1814, with a pension of 6d a day, Harris became a cobbler in Soho, London. Here he encountered a retired officer who was interested in Harris' military experiences and wrote them down. Harris was probably unable to read or write, but seems to have been remarkably articulate. His observations on the officers he met were as follows:

The officers are commented on and closely observed. The men are very proud of those who are brave in the field and kind and considerate to the soldiers under them. An act of kindness done by a officer has often during the battle been the cause of his life being saved. Nay, whatever folks may say upon the matter I know from experience that in *our* army the men like best to be officered by gentlemen, men whose education has rendered them

Royal Green Jackets Museum, Winchester. Here you can see a presentation of the battlefield of Waterloo.

more kind in manners than your coarse officer, sprung from obscure origin, and whose style is brutal and overbearing.

In the later stages of the Napoleonic Wars, which extended over nearly twenty years, the traditional source of officers became exhausted and there were promotions from the ranks, which changed the image of the officer caste. But Harris went on to say:

My observation has often led me to remark amongst men that those whose birth and station might reasonably have made them fastidious under hardship and toil have generally borne their miseries without a murmur, whilst those whose previous life, one would have thought, might have better prepared them for the toils of war have been the first to cry out and complain of their hard fate.

And let me bear testimony to the courage and endurance of that army under trials and hardships such as few armies at any age, I should think, endured. I have seen officers and men hobbling forward, with tears in their eyes from the misery of long miles, empty stomachs and ragged backs, without even shoes and stockings on their feet, and it was not a little that would bring a tear into the eyes of a Rifleman of the Peninsula.

Harris' experience was, of course, limited to his own campaign. There were many other occasions when armies, both British and foreign, would suffer similar or worse hardships than the ones he suffered.

THE STUART REBELLION

The ambitions of Louis XIV to dominate Europe and Britain were finally ended by the Treaty of Utrecht in 1713. Unlike most peace treaties which seem to contain the seeds of another war, that of Utrecht preserved a general peace for thirty years. However, the fact that there appeared to be no danger to the country after 1713 caused the government to run down the army to save money. Queen Anne died in 1714 and, as she left no children, the next heir to the throne was a Hanoverian, a grandson of the sister of Charles I (who had been executed after the English Civil War). His name was George, he was 54, spoke no English and did not intend to learn the language. For much of his reign (1714 to 1727) the country was ruled by the prime minister, the capable but cynical Robert Walpole. As George spoke no English and Walpole no German, their attempts at conversation were made in Latin, a language which neither of them knew well.

However, the fact that an obstinate and unintelligent German was now the king of England seemed to offer a golden opportunity to the Scots, who believed that the son of James II (who had fled to France when William III landed in England) could now replace him on the throne and rule as James III. The ensuing brief campaign was a disaster in which the Scottish leaders made a series of foolish mistakes and 'the Old Pretender', as James was dubbed, was soon back in exile. However, the Stuarts were not prepared to relinquish their claim to the English throne and they persuaded the Spanish to assist them; although the French refused to help. Consequently, a few hundred Spanish troops landed in Ross-shire in 1719. Although they were supported by some of the nearby clans, the planned insurrection was never a viable proposition and the rebel army was defeated at the battle of Glenshiel. Spain had declared war on England but soon afterwards signed a humiliating peace.

However, when the British Cabinet heard that Spaniards had landed in Scotland there was great alarm, particularly as the British army had been neglected in the seven years since the Treaty of Utrecht. The government was now so pressed for military resources that the Chelsea Pensioners, who became known as Fielding's Invalids, were turned into a fighting force which stayed in existence for seventy years. Doubtless many of the present Chelsea Pensioners would gratefully leap at a similar opportunity and make up in wisdom and experience what they lacked in mobility.

Despite the landing of the Spanish in Scotland, the British government did not learn its lesson of the folly in running down the strength of the army. At the end of Marlborough's campaigns, troop numbers had dropped from 200,000 to 30,000, and fell to 12,000 in 1719. Had the quality of the troops been good this fall in numbers might not have been so dangerous, but unfortunately the soldiers, their officers and the general attitude of the public towards them had seldom been at a lower ebb. Numbers were increased to 18,000 after Walpole became prime minister in 1721, but this figure was far too low to cope with Britain's various commitments.

THE WAR OF JENKINS' EAR

Walpole's aim was to avoid war, which he considered to be expensive and undesirable. Consequently, when sea captains brought complaints about Spaniards attacking British ships which were legally trading with South America, he chose to ignore them. However, in 1739 a sea captain named Jenkins claimed that not only had his merchant ship been boarded by Spaniards but they had even cut off his ear. To prove his point, he brought his ear back in a bottle and displayed it to parliament. Natural indignation boiled over and Walpole was first pressed by parliament to demand reparation and then, when this was not forthcoming, to declare war. The country as a whole rejoiced, for in the previous twenty years they had forgotten the hardships of war. When excited people rang the church bells in the general rejoicing, Walpole observed gloomily: 'They are ringing the church bells now but soon they will be wringing their hands.' He did not have to wait long to see his prophecy come true.

'The War of Jenkins' Ear', as it became known, was the preliminary to a series of wars which ranged from Europe to countries as far apart as India and North America. As would happen again in the future, the army emerged from years of neglect, bad food and poor pay, and won distinction wherever it was required to fight. Unfortunately, in the War of Jenkins' Ear, the enthusiasm of the soldiers was not matched by the competence of the senior officers, and an expedition against Cartagena, a Spanish port in South America, resulted in a repulse which was followed by an epidemic of yellow fever. The casualties from these two disasters were so high that the government was spurred into taking steps to prevent a repetition. 'Yellow Jack', as the deadly fever was called in the army and navy, was already familiar to soldiers who were serving in unhealthy foreign stations, but this occasion, when 3,000 men died in four days and 90 per cent of the entire expedition died in battle or from disease, was something new in their experience.

The disasters of Cartagena were almost matched by those of an expedition which had set off under Admiral Anson to harass the Spaniards on the other side of South America and disrupt their trade with the Far East. Anson's eight ships carried a total of 250 soldiers but, as these were all Chelsea Pensioners, most of whom were over 60 and some over 70 years old, it was hardly surprising that they were unable to withstand the conditions. Four-fifths of the men, and seven out of the eight ships, failed to survive, but Anson conducted a series of successful raids and captured an enormous treasure galleon.

The War of Jenkins' Ear soon merged into a European conflict

The Black Watch at Fontenoy 1745.

known as the War of Austrian Succession. When the Emperor of Austria died in 1740 he left his vast empire to his daughter, having previously persuaded the major powers in Europe to guarantee her succession. In the event, instead of protecting the young woman (Maria Theresa) they fell on her like wolves, attempting to seize portions of her territory. The Prussians were the first to move and the French and Spanish were quick to follow. England had no intention of allowing any one of them to seize the Austrian domains and thus to alter what became known as the 'Balance of Power' in Europe. The Balance of Power theory persists to this day: it operated in the Napoleonic Wars, it produced the combination of the Allies against Germany in World War I and again in World War II, and is now represented by the balance between the NATO and Warsaw Pact powers. England's interest was not merely to prevent a single power becoming dominant in Europe, but also to keep France so busy on the Continent that she would have little time or resources to spend on colonial ventures.

Walpole resigned as prime minister in 1742. His place was taken by Carteret, who promptly took 16,000 Hanoverians on the English pay-roll and also raised eleven new English regiments. These included the Oxfordshire and Buckinghamshire (now part of the Royal Green Jackets), the Essex, the Sherwood Foresters ⌐⌐ , the 2nd Duke of Cornwall's Light Infantry and the Northamptonshire Regiment (now part of the Royal Anglian). Although these military moves were mainly implemented as a precaution against French actions, England was not at war with France until 1743. However, when England became directly involved in war, George II, who had fought in Marlborough's army, took over command. Unlike his father, George I, George II could speak English, although he spoke it with such a thick accent that few could understand him. However, there was nothing wrong with his personal courage and at Dettingen he displayed considerable tactical skill as well. A story is told of him that his horse, having been scared by gunfire, bolted at the beginning of the battle, so he dismounted at the earliest opportunity and from then onwards fought on foot. He said, with truth: 'I can trust my own legs not to run away with me.' Two years later, George II's son, the Duke of Cumberland, distinguished himself in the battle of Fontenoy (which the British lost) although he was only 25 years old.

BONNIE PRINCE CHARLIE

The British government found itself with more immediate troubles than a continental war when, on 5 April 1745, Charles Edward Stuart, 'Bonnie Prince Charlie' – 'the Young Pretender' – landed in Scotland with the intention of succeeding where his father, 'the Old Pretender', had failed. Charles was different in character from his morose father and soon attracted widespread support. The government in England was hard pressed to find an army to confront him, as the bulk of the army was engaged overseas in the war with

Essex Museum, Oaklands Park, Chelmsford. Sherwood Foresters Museum in Nottingham Castle. In both these museums eighteenth-century weapons, uniforms, paintings, equipment and medals can be seen.

France, which was not going well. Charles won a brisk victory at Prestonpans, near Edinburgh, and pushed on towards London. But London is a long way from Edinburgh and many of Charles' supporters, who came from remote parts of the Highlands, decided that it would be a good time to go home with the spoils they had collected. When Charles reached Derby he found his army much depleted by these temporary desertions and was persuaded to fall back to Scotland and wait for the French to give him support. He was well aware that the Duke of Cumberland was now on his way back from the continent with battle-hardened troops. But the French failed to support Charles and after a final victory at Falkirk, he met total disaster at Culloden Moor on 28 January 1746.

Charles had proved an inept general and his army was half-starving when it met Cumberland's well-supplied forces. He escaped from the battlefield and hid in the Scottish countryside until he could find a boat to take him back to France. Although there was a price of £30,000 on his head, not one of his countrymen dreamt of betraying him. Cumberland's troops hunted down the surviving Scots with relentless barbarity. One result of the war was that a road was built into the Highlands, and although intended for military purposes, it also opened up the country commercially.

Culloden, Inverness.
Culloden is a well-preserved battlefield, now a National Trust property. A notable feature of it is that heather has never grown over the Highlanders' graves. Charles' ADC was the chief of the Maclachlan clan; he lies in the centre of the front line of the graves.

The Battle of Culloden Moor was the end of Stuart attempts to regain the throne, but no cause could have ended with more determination and courage. Hopelessly outgunned and outnumbered, the Scots fought on, many with several wounds, some even with a severed limb; they almost achieved the impossible. The last flicker of hope of the Jacobites was extinguished when the male line of the Stuarts died out in 1807.

FAR-DISTANT CAMPAIGNS

The war in Europe drifted to an end in 1747 and a peace treaty was signed in 1748. It was realised by all concerned that this peace treaty merely signified a breathing space to the exhausted combatants. Another war was already on the horizon and would be even greater and more wide-ranging.

However, even the War of Austrian Succession had not been confined to Europe and in campaigns in far-distant territories the British soldier proved how adaptable he could be, whatever the terrain or climate. The war did not reach Canada until 1746 and, because of the slowness of communication, was still continuing long after the peace treaty had been signed. One of the most frustrating aspects of fighting in countries so far distant that it took a ship – and thus a message – several weeks to reach them, was that armies could be fighting desperate battles to win territory which a peace treaty, signed months earlier, could already have awarded to one side or the other. There were bizarre situations in India where, for example, English and French officials were happily entertaining each other when they learnt that they had already been at war in Europe for several months.

The Battle of Prestonpans 1745, won by 'Bonnie Prince Charlie'.

Up to this time, the French had been very successful explorers in North America, where their presence is traceable from place-names like New Orleans, Louisiana and Quebec. Their pioneers made friends with Red Indian tribes whom they persuaded to be pro-French and anti-British. The French established a strategic fort halfway between Erie and the Ohio river and called it Fort Duquesne. The British subsequently captured it and named it Fort Pitt; it is now Pittsburgh (named after the British Prime Minister, William Pitt).

Until the eighteenth century, much of North America was still unexplored and neither Canada nor America existed as a nation. British settlers were in a strong position in the eastern portion of what is now the United States of America, but in the north the French had strong reasons for supposing that they could turn the country into a French colony. The early settlements in overseas territories had been made by chartered trading companies. In order to settle and trade in a newly discovered territory a government charter was required; this was at first granted to single ships, then later to settlements. The government could then tax and regulate trade and in return would protect the settlers from attacks by rivals. The early settlers in America had gone out with the Virginia Company to trade in timber and tobacco, but the future Canadians had

Troopers; Light Dragoons 1742.

been members of the Hudson's Bay Company, whose business was fur and fish.

The East India Company operated commercially in India and for many years employed its own soldiers. Although Canada and India are vast countries in which there should have been room for all commercial ventures, neither the home government nor the early settlers considered it to be the case. In India there was brisk

Cavalry guidons.

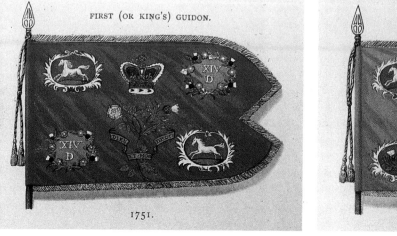

FIRST (OR KING'S) GUIDON.

1751.

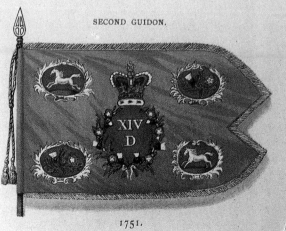

SECOND GUIDON.

1751.

fighting during the War of Austrian Succession. The French captured Madras, but the English besieged the French possession of Pondicherry. While the fighting continued, the combatants learnt that as part of the treaty which had ended the war Madras would be returned to the British and Louisbourg (in Canada), which the British had captured, would be handed back to the French.

However, during the next ten years, although Britain and France were officially at peace there was a series of bloody clashes between British and French troops in both America and India.

The outstanding British general in India during this 'peace' was Robert Clive. ☞ Clive had originally been a clerk in the East India Company's service, but he found that a military life suited him better. In 1751, although he was only a captain, he found himself the senior British officer when the French were establishing a dominant position based on the town of Arcot. With an army consisting of 200 British and 300 locally raised Indian troops, he marched 65 miles (105km) in five days and captured Arcot without difficulty. Marching an average of 13 miles (21km) a day for five days and fighting a battle at the end of it would be a hard task, even in a temperate zone, but in the exhausting heat and harsh conditions of India, it might well have seemed a near impossibility. However, Clive was one of those leaders whom Britain has always managed to discover in an emergency. Subsequently, Clive held the town against a siege by 10,000 French and Indian troops. After fifty days his attackers withdrew. These victories were achieved with East India Company troops, which a hundred years later became part of the Indian Army.

The first British regular infantry regiment arrived in India in 1754 and was accompanied by some Royal Artillery. The infantry regiment was the 39th Foot, for in 1751 all British regiments had been given numbers (instead of bearing the names of their colonels). Later, the 39th Foot became the 1st Dorsets and are now part of the Devon and Dorset Regiment.

Powis Castle, Wales. The relics of Robert Clive are housed here.

THE ROYAL ARTILLERY

By 1720 it was clear that the development of new types of gun required a regiment of specialists. Consequently, in 1722 the Royal Regiment of Artillery was founded with four companies – that is, about 500 men. Gunnery was clearly a technical matter and to provide the new regiment with officers, the Royal Military Academy was founded near Woolwich in 1741. Woolwich was already the site of the Royal Arsenal. Although the army did not regard the establishment of the new regiment of artillery with great interest and chose a Dane for its first commanding officer, it was noted that a more efficient system than purchase was needed if its officers were going to achieve the technical standards required for the proper handling of new weaponry. Consequently, neither commissions nor promotion in the Royal Artillery would be obtainable by purchase; technical and educational merit and meritorious service would be the only criterion. Needless to say, the fact that commissions in the Royal Artillery could be obtained by merit brought a flood of applicants to the new arm. Other branches of the army tended to regard 'clever' officers with some suspicion and to look down on them, but the 'Gunners', as the Royal Artillery were proud to call themselves, soon earned the respect of their comrades, not merely for their technological skills but for their courage and ability in the other military arts. The fact that guns required horses and the handling of horses demanded a high degree of skill, earned the grudging admiration of other units; it was a British tradition to judge a man's merit by his skill at riding and controlling horses.

The RMA, Woolwich, continued to flourish and earn distinction, having on its staff notable military or civilian instructors, one of whom, Michael Faraday, was a physicist, chemist and the discoverer of electromagnetic induction. In 1947, Woolwich merged with the Royal Military College, Sandhurst, which had not been founded until 1812, and the new establishment thenceforward became the Royal Military Academy, Sandhurst.

Thus, by the mid-eighteenth century, the British army had acquired many of the assets that were necessary if it was to become an efficient fighting force. There was still much to be done, however. Engineering studies were not properly organised and the Corps of Engineers would not be established until 1786. However, the first Scottish regiments were now incorporated in the British army. In 1725 four companies of Highlanders had been raised and wore a very dark tartan. Fourteen years later, four more companies were added and the regiment was given the name of 'The Black Watch'. The Royal Inniskilling Fusiliers was raised in 1688 and has since fought all over the world. They held the centre against the French at Waterloo when defeat seemed probable.

Sandhurst, Camberley. Although modern security regulations have made access to Sandhurst more difficult, it is possible to visit it once a year on the combined Open Day and Horse Show, and also at other times if written permission is obtained.

Balhousie Castle, near Perth. Here you will find an excellent museum containing exhibits from the history of the Black Watch regiment.

FOOD, HOUSING AND CLOTHING

During the eighteenth century there were very few permanent barracks which meant that most of the army had to be billeted on an unwilling civilian population. Fortunately, they did not have to be quartered in private houses but were accommodated in inns or the attached stables. Needless to say, they were not given comfortable rooms and beds in these establishments, but merely a roof over their heads, stabling for their horses, water and materials to make a fire, and pots and pans for cooking. No cooked food was provided – that was the soldier's own responsibility. Bedding was straw. Although straw makes good bedding for animals, it has its limitations when it is used for humans as it becomes musty and insect-ridden. Nevertheless, the army persisted with straw palliases up to the middle of the present century. At the start of World War II recruits found themselves issued with three 'biscuits'. 'Biscuits' were square straw palliases and three biscuits made up the length of a bed. Sometimes the straw was all in one mummy-shaped container which was hardly wide enough for the average soldier to lie

An officer of the 42nd (Royal Highland Regiment) in 1792.

A powder horn made in America during the Seven Years War (1756–1763) and carried by a soldier of the 42nd (Royal Highland) Regiment from a strap decorated with Huron Indian bead and moosehair embroidery.

on. But at least the 1939 soldier was in dry accommodation, which was more than could be said for his eighteenth-century predecessor, whose only protection against rain was a leaking roof.

Cooking facilities in the mid-eighteenth century meant the use of the kitchen – with perhaps two containers – when everyone else had finished with it. As the soldier's ration was gristly meat and potatoes, there was not much scope for culinary skills. The innkeepers detested having soldiers on the premises, for the military were not ideal guests. Although they had little left out of their meagre pay after compulsory and incomprehensible stoppages (which included the cost of their scanty rations), it still left just enough for them to make themselves drunk – for example, 2d out of their daily pay of 8d. The rations which they had paid for provided two meals: breakfast between 7am and 8am and dinner at midday. However, each man had a standard issue of 5 pints of beer a day. Most of this would be drunk in the nineteen hours between dinner and breakfast, but the 2d the soldier had in hand enabled him to buy spirits as well. Gin was cheap in the inns, hence the saying 'drunk for a penny, dead drunk for tuppence', but most soldiers found cheaper and even more potent spirits in the canteens (grog shops), which were run inside the camps by civilian contractors, with official approval. Doubtless the aim was to see that most of the profits from soldiers' drinking habits found their way back to the regiment, but this seems to have been an optimistic view. Even in the present century, soldiers have stories of the ability of local contractors to make a profit on the narrowest of margins: the Chinese were said to have had a special genius in obtaining more cups of tea from a packet than anyone else would have thought remotely possible; this did nothing for the taste, according to the soldiers who bought it.

The fact that the second and last meal of the day was over by 1pm did not mean that the soldier was then free for the day. On active service he was lucky if he had any free time at all, but then his feeding arrangements were better. In peacetime at home he would have been up since 6am, drilling if he was an infantryman and tending his horses if he was a cavalryman or a gunner. The mornings were occupied by a combination of drill, weapon training and inspections, and the afternoon with what the army, with some accuracy, describes as 'fatigues'. An ingenious NCO can always find some task which occupies a body of men, but after the afternoon chores the rest of the day was, in theory, the soldier's own.

Unfortunately for the soldier, part of his 'spare' time was spent on carefully maintaining his appearance. George II was devoted to the army and, when he was commanding it, had ordained that the soldiers' uniform had to be made dashing and attractive. He ordered that the brims of hats must be turned up on three sides; this undoubtedly made them look smarter, but at the same time made them useless as a protection against rain. Hair had to be drawn back tightly on the head and made into a pigtail at the nape of the neck; the whole head was then powdered. Added to this time-consuming exercise was the polishing of all a soldier's buttons and the pipe-claying of his breeches and gaiters. Gaiters had replaced the practical woollen stockings because George II thought they looked smarter; they were at first called 'spatter dashes', a name which became shortened to 'spats'. (Until the 1930s elegantly dressed civilians wore short spats over their shoes; it was said that they were for keeping the feet warm, but in fact it was a mark of elegance.) George II also altered the practical cut of the soldier's coat, which was shortened and given very wide facings. Off-duty soldiers would turn their coats inside-out and wear them that way to prevent them from becoming dirty. By the time George II had finished introducing modifications, the soldier's uniform was picturesque but not practical; nevertheless, it was what he had to wear, whether he fought in the bleak conditions of Canada or on the burning plains of India. Impractical though this uniform was, however, there was worse to come.

In the middle of the eighteenth century the government decided that it would be better for the army if soldiers were accommodated in barracks, rather than in shacks behind alehouses, and during the following fifty years a building programme ensured that some two hundred barracks were built. The definition of a barracks seems to have been a building with a roof on it. There was no heating and no facility for washing – the smell in winter must have been appalling. Each barrack-room was supplied with a tub which had the dual function of being used as a urinal at night and a washtub by day, but was used rather less for the latter purpose than the former.

Although marriage was discouraged, the army accepted a limited number of women on the strength that they could help with the chores when other duties prevented the soldiers from tackling them. Six wives were permitted among every hundred men. They were allowed free rations but were expected to mend clothes, cook and clean. Their accommodation was free, too; their bed space was shielded by blankets which were slung across the end of the barrack-room, the rest of which was occupied by unmarried soldiers. If an unmarried soldier decided to marry 'off the strength' – that is, without official approval and without permission, his wife was not allowed to sleep in the barracks, nor he out of it. Fortunately, the women who married soldiers were usually capable of looking after themselves and their husbands, too.

Conditions for officers' wives were vastly better than for those of the lower ranks, but even officers were not allowed to marry without their colonel's permission, which was given reluctantly. Until well into the present century, officers were discouraged from marrying young by having their marriage allowance withheld until they were 25 years old. Officers without substantial private means usually preferred to wait until they had reached the rank of major; at that point they felt that they could support a wife in the style to which she was accustomed. This often meant that the wife was much younger than her husband. Any children of officers serving abroad were sent back to England for education at an early age, and because of travelling difficulties this often meant that children did not see their parents for a dozen or more years. Rudyard Kipling has much to say of this practice and the misery it caused to all concerned because he had experienced it himself.

Like many buildings that were subsequently provided for the use of the services, the eighteenth-century barracks were built rapidly in the belief that they would soon be replaced by something better. In fact, they and their successors stayed in use for nearly two centuries, in spite of the fact that periodically they were condemned as unfit for habitation. There is a replica of a 1900 barrack-room in the Aldershot Military Museum; such barrack-rooms were still in use in 1939 and were preferable to much of the flimsy accommo-

Aldershot Military Museum, Aldershot. There is a replica of a 1900 barrack-room here. Such barrack-rooms were still in use in 1939 and were preferable to much of the flimsy accommodation which was hastily provided for the wartime intake of recruits.

dation which was hastily provided for the wartime intake of recruits.

In theory, regiments were posted abroad and brought home after their tours of duty – usually much depleted from disease or even desertion – but sometimes they were forgotten. For example, the 1st Staffordshire Regiment was sent to the West Indies in 1706 and was not brought home for sixty years; by that time all the original soldiers had been replaced. 🔫

Whittington Barracks, Lichfield. War relics of the Staffordshire regiment can be seen in the museum here.

Castle Museum, Castle Barracks, Enniskillen. Uniforms, colours, battle trophies, bugles, Victoria Crosses and a host of other exhibits can be seen here.

MILITARY CUSTOMS

Many military customs, which had originally been adopted from other countries, were now firmly established in the British army. Military music is based on a combination of the oriental drum and the European pipe (the pfeif – hence drum and fife bands). 🔫 The colours, which no longer appear on battlefields (but are very popular on regimental ties), were originally rallying points where a soldier knew he would find his friends. It was a great honour to carry the colours, to be the standard-bearer, and the holder of that office would not hesitate to sacrifice his life to preserve the sacred regimental flag. To capture an enemy's standard was a feat of great distinction and was honoured accordingly. Unfortunately, in modern warfare it is important not to disclose the identity of individual units in case the enemy gains an advantage from that knowledge, so a modern soldier goes into battle with all the more obvious marks of identity removed. If he is captured, a soldier must never reveal more than his name, rank and number. Intelligence officers, knowing that prisoners have been ordered not to disclose the name of their unit, resort to various tricks to try to bluff them into revealing it.

One of the most revered British customs is that of firing three volleys over a soldier's grave. British it may be now, but it was German in origin. Ranks, too, have an international flavour, although their significance may vary between countries. The rank of sergeant originally had a much higher status than it does today. Corporal, the first promotion from the basic rank of private, is a rank which has been held by such notorious figures as Hitler and Napoleon. Lieutenant (place-holder) is an obvious French derivation but, curiously enough, no one seems to know the origin of the word 'colonel'. For many years colonel was spelt 'coronel' and this is approximately the modern pronunciation of the word. Many people are puzzled by the fact that the modern ranks of general go from major-general, to lieutenant-general, and then to general, although a major is clearly a higher rank than a lieutenant. The reason is that major-general was originally sergeant-major general, but the 'sergeant' was gradually dropped. There was formerly a rank of 'captain-general', signifying a commander-in-chief, but this has long since been abandoned. The title of 'field-marshal' is an honour rather than a rank, for it can be held by people without military experience, such as members of the royal family.

CHAPTER 4

A WORLDWIDE WAR AND ITS AFTERMATH

THE SEVEN YEARS' WAR

The Seven Years' War which broke out in 1756 had the perils and characteristics which later became closely associated with other conflicts, including those of World War II. They included initial disasters, atrocities, the threat of the invasion of England, and eventually superhuman efforts by the armed forces which led to final victory. The war saw the emergence of a great statesman who rallied the country as Churchill did in World War II, and it brought fame to two of the most brilliant but unconventional generals in British military history.

Although the Seven Years' War was essentially a war between England and France, with their supporting allies, in practice it became three separate wars. The peace treaty which had ended the previous war in Europe (the War of the Austrian Succession) had been largely ignored in India and Canada, and even in Europe it was regarded as little more than a temporary ceasefire between the combatants. The French had decided that the English, who had frustrated their plans earlier in the century, must be dealt with once and for all and that the most effective way of doing this would be an invasion. To that end they decided to send 50,000 men across the Channel in flat-bottomed boats. The French were aware that if the British navy was in the vicinity when the French invasion fleet sailed they would soon be attacked, so they planned their departure to coincide with a time when the British navy was occupied elsewhere. The French therefore planned to attack Minorca, which was an easier target, to draw the English off. However, the British had learnt of the French invasion plan from secret agents and decided that the planned attack on Minorca was a decoy. Consequently, she concentrated most of her ships in the Channel and left Minorca virtually undefended. This gave the French their first victory.

Having lost Minorca, Britain was stunned by the horrific news of what became known as 'The Black Hole of Calcutta'. Calcutta was a trading post, not a military station, so there were very few soldiers in the town. However, when a treacherous native prince (the Nawab of Bengal) decided to attack Calcutta, the makeshift garrison put up a strong resistance for three days before it was overwhelmed by superior numbers. The 146 survivors were then induced to surrender by the offer of fair terms, but, having done so, the entire group was crammed into a cell 20ft (6m) square. As there was practically no ventilation and Calcutta in June is stiflingly hot, only twenty-three of the captives were alive the following morning. When the news of the two disasters that had occurred at Minorca and Calcutta reached England, there was a public outcry followed by a change of government. William Pitt became prime minister and soon roused the country into pursuing the war actively. Clive, who had become commander of the British forces in India, now disposed of the Nawab's forces at the Battle of Plassey. Clive had only 3,000 men to put against the Nawab's 50,000, and to reach Plassey his army had to march through monsoon rains in temperatures up to 90°F (32°C). The ensuing battle was won by morale rather than by gunfire, for Clive's force advanced with such determination that the Nawab's men were reluctant to fight. Casualties were low on both sides. The Nawab fled from the scene early in the battle, but was later strangled by one of his many enemies.

But even with the evil Nawab despatched, the task for British troops in India was by no means over. By now the French had reinforced their garrisons in India and Clive had considerable difficulty in capturing their strongholds. Unfortunately, Clive's reputation was impaired by the accusation that he had accepted presents from grateful native rulers. Clive could not deny it, but said that the sums were small and, in view of the huge gifts he had been offered, he was 'astonished at his own moderation'. However, the British parliament has always had a keen nose for presents which might involve bribery, and Clive was not the last officer to be accused of corruption. Three hundred years earlier, a successful general would have been commended for obtaining large rewards for himself and everyone else, but those days were long past.

One of Pitt's first moves had been to introduce compulsory military service and, with this, fifteen new battalions were raised. Some of them were second battalions to existing regiments, but others were new; they included the Durham Light Infantry [image], the Highland Light Infantry and the Seaforth Highlanders. After long and distinguished careers, the Durham Light Infantry (DLI) has now become part of 'The Light Infantry', the Seaforths have merged with the Cameronians to become 'The Queen's Own Highlanders' and the Highland Light Infantry joined with the Royal Scots Fusiliers to become The Royal Highland Fusiliers.

During the Seven Years' War the brigade system was introduced by which regiments were grouped into fours (now threes). This organisational change meant better arrangements for the distribution of ammunition, food and clothing. During the war officers were promoted for efficiency, not by purchase, although purchase was reintroduced subsequently.

In Canada the French were in possession of the Great Lakes area. In 1758 a British attempt to advance on the French stronghold of Montreal was defeated with the loss of 2,000 casualties; 500 of these came from the 42nd Regiment, the renowned 'Black Watch'. However, even though the attempt to capture Montreal failed, the force had some consolation in the capture of Frontenac. Frontenac is now Kingston and the site of the Canadian Royal Military College.

Important though Montreal was, it took second place to Quebec, whose enormous strategic value lay in its position where the St

Durham Light Infantry Museum, Aykley Heads, Durham. Here you will find regimental relics, a 17 and a 2 pounder gun, and a Bren gun carrier.

Lawrence narrowed. The task of capturing Quebec fell to General James Wolfe, who seemed a most unlikely person to have become a soldier. He was undersized, unimpressive and sickly (but not as sickly as he imagined himself to be), and was deathly pale; nevertheless, he had astonishing powers of endurance when the occasion demanded. At this time he was thirty-two but had been fighting since he was sixteen and at Dettingen had acquitted himself well. A memorable remark was made by George II, when it was suggested that Wolfe was mad. 'Mad is he?' replied George. 'Well, I wish to God he'd bite some of my other generals and make them mad too!'

Wolfe set off with 8,000 men, but disease and casualties soon reduced this number to 5,000. The aim of the expedition was to storm the Heights of Abraham, on the top of which Quebec was situated. First, however, the fleet that was carrying the expedition had to find its way up the tricky waters of the St Lawrence, against considerable French opposition: French guns were well sited against such a contingency. Fortunately, the navigator of the English expedition was Lieutenant James Cook, who had worked his way up from the position of ordinary seaman and who would later, as Captain Cook, earn fame as a Pacific explorer. The French tried to destroy the English boats with fireships but Wolfe was not easily deterred; the fact that the French were reported to have 7,000 men on the heights simply spurred him on.

Whitby, North Yorkshire. The town of Whitby, of which Captain Cook was a native, has many relics of Cook.

Success was achieved when Cook divided the expedition into two halves, one occupying the attention of the French, while the other quietly slipped upstream to a point 2½ miles (4km) up-river from Quebec. The soldiers then ascended by a narrow path in the dark and, when challenged by a French sentry, replied with convincing answers and accents. Wolfe's army reached the Plains of Abraham and was advancing cautiously when it was attacked by Montcalm's French troops. This was a rash move by the French commander, who would have been wiser to have let the British try to attack defended positions, but Montcalm's annoyance at discovering that the British had advanced so far, seems to have distorted his judgement. The French troops hurled themselves towards the Redcoats, who held their fire until the last moment, although they lost many men in the process. At 35yd (32m) 'when they could see the whites of the French eyes' they could not miss, and their successive volleys devastated the French ranks. Wolfe then ordered the British to advance forward and he himself charged with the Gloucesters. He was wounded three times, the last time fatally, but was still giving orders with his dying breath. Montcalm was also killed in this battle.

Westerham, Kent. Wolfe was born at Westerham in Kent, where his former house contains relics of his career. Robert Clive was born in Market Drayton in Shropshire but was educated at the Merchant Taylors School in London and retired to Powis Castle in Wales; the latter has many interesting relics of him.

The Seven Years' War began in Europe even more disastrously than it had done overseas. The Duke of Cumberland, who was nicknamed 'the butcher of Culloden' because of the atrocities committed by his troops on the battlefield, had made a disastrous start on the Continent and been heavily defeated by the French at Hastenbeck. Soon afterwards he was replaced by another general,

Quebec House, Westerham. Here you will find an exhibition of the Battle of Quebec. Squerryes Court, Westerham, also contains a 'Wolfe Room'.

Ferdinand of Brunswick, and another army which met the French at Minden, where nearly 100,000 men clashed on the battlefield. The British infantry, who were in the centre of the battle formation, mistakenly advanced before they were expected to do so. This move brought them to close quarters with the French cavalry, on whom they inflicted heavy casualties. The battle is memorable as an occasion when infantry charged cavalry and routed it, and the supporting gunners used their horses to charge the cavalry.

The British cavalry were badly commanded at Minden and did little to win the battle, but they made up for this the following year (1760) at Emsdorff, where the 15th Hussars (now the 15th/19th) charged with such skill and ferocity that they chased the French for 20 miles (32km). The cavalry distinguished itself again at Warburg, where the French commander mishandled his troops and left 20,000 men in an exposed position. This presented a golden opportunity to the British cavalry, as long as they could cover the distance of 5 miles (8km) before the French had time to escape. They were led by the Marquis of Granby, whose hat and wig fell off in the confusion of the charge. Granby's bald head shone as brightly as a beacon in the sunshine, and the origin of the expression 'going bald-headed for victory' lies in this incident. The phrase is nowadays used by racing and other sporting commentators, who probably do not suspect its origin.

The Marquis of Granby was a very popular commander who was always attentive to the welfare of his men. He was criticised by his fellow commanders for his leniency towards offenders but he ignored their criticism. After the war he helped many of his soldiers, mainly NCOs, to open inns and they, in their turn, commemorated his name by putting it on their inn signs. Consequently, there are many inns named the Marquis of Granby around Britain.

Granby's own regiment was the Blues (now the Blues and Royals) who won great distinction at Warburg. In one of the later battles of the European campaign the Northumberland Fusiliers (the Fighting Fifth), who are now part of the Royal Regiment of Fusiliers, helped Granby to win another spectacular victory.

The Seven Years' War ended in 1763 after a number of sea battles in which skilful naval commanders such as Hawke, Anson, Boscawen, Saunders and Watson played a distinguished part.

Pitt, who could rightly be described as the architect of the wartime victory, was dismissed from his office of prime minister before the war had ended because George III, who ascended the throne in 1760, resented having such a powerful prime minister. Both George I and George II had been content to let the prime minister govern the country while they pursued their own not very reputable activities, but George III had different ideas. His folly and stubbornness subsequently led Britain into war with the American colonies and the disasters of the War of American Independence.

The Seven Years' War ended with large gains for Britain and was subsequently described as 'the war which established the British

Combermere Barracks, Windsor. The museum here holds many relics of what was then the Blues (now the Blues and Royals).

Cavalry was not merely impressive in appearance, it was also very good in battle.

Empire'. In the peace treaty (the Treaty of Paris) France relinquished her claims in Canada, but was allowed to retain her settlements west of the Mississippi. Minorca was returned to Britain but the French retained their former possessions in India, provided they did not fortify them. Britain received Florida, but Havana and Manila, both of which had been captured in the war, were returned to Spain (who had been France's ally).

Once the war was over the British public soon reverted to its former attitude towards soldiers. This is aptly described by Kipling in the lines:

> Oh it's Tommy this and Tommy that and 'Get away you brute'
> But it's forward, Mr Atkins, when the guns begin to shoot.

The public's attitude to wartime leaders has also had a certain consistency throughout history. Pitt was hounded from office after leading the country from great peril to overwhelming victory; Clive, one of the great generals of the Seven Years' War, having been blamed for matters which were outside his control, eventually committed suicide. After losing his post as prime minister, Pitt remained out of favour for years and was only returned when once more the country faced an emergency; by that time, however, he was too old to make an effective contribution. In 1940 Winston Churchill rallied the country with stirring speeches and gave the British people the belief that Britain could overcome what seemed overwhelming odds; in 1945 he and his government were swept out of office by an electorate which seemed to have forgotten his great contribution so readily. Like Pitt, he was removed from office while the war which he had done so much to win still continued.

THE WAR OF AMERICAN INDEPENDENCE

The reign of George III (1760 to 1820) was a catalogue of disasters which caused much unnecessary suffering to British soldiers. During his last ten years George III was hopelessly insane and the country was governed by a regency. George's determination to be ruler as well as king led him to surround himself with incompetent advisers who would flatter but not criticise him. The major event of his reign was the loss of the American colonies and the founding of the United States of America. Although the establishment of an independent America was undoubtedly an inevitable and desirable development, the manner in which it occurred caused considerable bitterness on both sides. To this day there persists in some parts of America a strong distrust of British motives. During World War II this showed itself in the often-voiced view that America should do nothing to help Britain to restore herself to her former position as head of the British Commonwealth and Empire. In recent years, of course, American understanding of British problems has developed considerably.

In 1774 there were thirteen separate colonial groups in America, which stretched from Massachusetts to Virginia. They were prosperous and growing, and, with the threat of a French attack removed, they felt independent and secure. Although they were confident that they could defend themselves, British troops had to support them to drive out forces led by an Indian chief named Pontiac who invaded the colonies. Uncertain that the colonists could in reality defend themselves, the British government decided that they should pay a modest tax as a contribution to the cost of maintaining British troops in America. However, the tax, which took the form of a stamp on legal documents, was so bitterly resented that it was soon cancelled. The American colonists objected to paying the tax because they did not have representatives in the British parliament. If the British government had conceded the point, the absurdity of trying to implement it would soon have defused the situation, but instead other taxes were imposed on such imports as tea, paper and glass; these also were soon cancelled. In America there were sharp divisions of opinion about the matter and many members of parliament also supported the American view. The upshot was that in 1775 a British force under General Gage was sent to destroy the military stores that the colonists were collecting at Concorde. The army was attacked on route to Concorde as well as on the return journey, and Gage himself was besieged in Boston. He then dispersed the colonists by capturing Bunker's Hill, but with heavy losses.

The Americans decided to defend themselves against the British and chose as their commander-in-chief a Virginian planter, George

Washington, who had fought for the British in the Seven Years' War. 🔫

The war was so unpopular in Britain that it was impossible to raise more than 25,000 men to send overseas. As this was insufficient for the task they had been set, their strength was made up by a contingent of Hanoverians and Hessians. There was some justification for sending the Hanoverians, for George III was still Elector (Prince) of Hanover as well as being King of England, but the Hessians were simply mercenaries. Pitt was now an old and very sick man, but he struggled to the House of Commons and vehemently denounced the use of foreign troops (or even of the Red Indians who it had been suggested might be used) against one's own countrymen. Pitt's speech put a great strain on his frail constitution, and he died soon afterwards.

The War of American Independence lasted for six years during which England found herself fighting not only the American colonists but also France, Spain and Holland, all of whom took advantage of the difficult position in which England now found herself. However, it is worth recalling that, just as the British were not unanimously in favour of the war, large numbers of American settlers also deplored it. This feeling caused 40,000 American settlers to join the British side and many of them emigrated to Canada when America won her independence.

The decisive moment of the war came when the British commander, Cornwallis, made a series of tactical mistakes and was besieged in Yorktown. With twenty-eight French ships cruising offshore, it was not possible for the British army to relieve him, and after a month Cornwallis surrendered with 4,000 men. Although a British force was still holding New York, the British government decided that continuing the war was futile and accordingly they recognised the declaration of independence that the Americans had proclaimed seven years earlier.

The war has many interesting features, and in some ways resembled the conflict which took place in South Africa 120 years later. The Americans, knowing that they were no match for the British troops in a straightforward battle, took good care to ensure that they fought on ground of their own choosing whenever possible; when Burgoyne's starving army was defeated at Saratoga, it was outnumbered by three to one. Overall, the British troops were totally ill-equipped for a war of this type. Their clothing was hopelessly unsuitable and their tactics were far too rigid. They were nicknamed 'the bloodybacks' by the Americans, not because of their red coats but because of the hideous injuries inflicted on them by the cat-o'-nine-tails.

The other victors of the War of American Independence were the French, who supplied both troops and stores to the colonists. Their major triumph was, however, at sea where they used good tactics and improved methods of gunnery. When the naval war began the Royal Navy was in a parlous state, having been neglected since the

Sulgrave Manor, Northamptonshire. George Washington's ancestors lived at Sulgrave Manor, which is now open to the public.

end of the Seven Years' War – for example, when a fleet of ships was sent to America, their masts broke in mid-Atlantic. Towards the end of the war the English were unable to relieve Yorktown because the French fleet that was guarding the coast was superior in numbers, quality and tactical ability.

During the war, Minorca had been lost again but Gibraltar held out, not without great difficulty. The commander at Gibraltar, General Eliot, was 59, and forty-three years of his life had been spent in the army (he had been wounded at Dettingen). Although he had a limited quantity of artillery, he made the best use of it, experimenting with red-hot shot and air bursts.

PEACE AND ITS AFTERMATH

There have been few peace treaties which have left everyone so dissatisfied. The Spanish recovered Florida and Minorca, but were left with a greatly strengthened Gibraltar on their doorstep. The Dutch received nothing for their efforts. The Americans, although they were pleased to have achieved their independence, resented the fact that those among them who had fought for the British were now to be settled in Canada to the west of Quebec. Britain felt that she had lost the war in spite of the fact that she had won most of the land battles and that the French were the principal cause of her misfortunes. The British were less than pleased with the Irish who had supplied the French fleet with provisions, but, still worse, when the British government had decided that the Irish Protestants should raise a volunteer force to protect the country from a foreign invasion, the volunteers noted that they were the only military force in the country and promptly demanded Home Rule. Britain had no option but to grant it and Ireland had Home Rule until 1800.

One positive gain for the British army was the emergence of the light infantryman. Previously, soldiers had been taught to advance in line and remain steady under fire; when their bullets were spent they would charge. But now certain commanders, one of whom was the future General Sir John Moore, realised that the mobile, quick-firing marksman would become an important asset. This did not mean that the other soldiers would become obsolete – on the contrary, they would win many more victories, but in future the British army would have a new flexibility. Among its proponents would be the 60th and 95th Rifles. The former were mentioned earlier as being a British regiment commanded by a Swiss colonel. It had originally been called 'The Royal Americans' and in 1758 had been issued with sixteen muzzle-loading rifles which, although they were more complicated to load, were more accurate to fire. The 95th did not come into existence until 1801, when it began as an experimental corps of riflemen who had been chosen to try out the new Baker rifle. The 95th, which would win great distinction in the Peninsular War, later became the Rifle Brigade and today, with the 60th Rifles and the Oxfordshire and Buckinghamshire Light Infantry, is part of the Royal Green Jackets. 🔫

Winchester. The headquarters of the Buckinghamshire Light Infantry is now at Winchester, where the museum has many interesting exhibits on display; these include medals from the Peninsular and Waterloo campaigns.

CHAPTER 5

THE NAPOLEONIC WARS

In 1783 Britain appointed William Pitt, son of the elder Pitt, as prime minister, at the age of only 24. Even at that period, his extreme youth drew some derisive comment from other members of parliament and it was said that the only reason he had his high office was the reputation of his illustrious father. In fact, William Pitt the Younger, as he came to be called, was entirely capable of discharging his numerous duties. He retained the premiership until 1801 and, after losing it for three years, was reappointed in 1804 for another two years. His twenty years in office coincided with the most dangerous period in British history prior to 1940. He became prime minister when the national finances were at their lowest ebb, but presided over a period of unprecedented financial and industrial expansion. However, this was a time of great political ferment and without Pitt at the head of the government, the French Revolution might have affected Britain more deeply than it did.

Wellington's Orders, with the Order of the Golden Fleece in the centre.

In 1789 there began in France a revolution which would eventually lead not to liberty, equality and fraternity, as the revolutionaries claimed it would, but to a period of terror and mass-murder which then gave way to a period of military dictatorship. Although the British government had at first tried to distance itself politically from the events on the other side of the Channel, there was growing revulsion against their bloodiness. Eventually, when the French royal family was executed in 1793, Britain could stand aside no longer. The war lasted, with two short intervals, until 1815, a period of twenty-two years. This century Europe has felt the strain of two wars, 4 and 6 years respectively, but the thought of over twenty years of warfare almost defies imagination.

BRITAIN UNDER THREAT

At the outbreak of war, France mustered 500,000 raw but enthusiastic troops; England had 30,000. This disadvantage was balanced by the fact that the Royal Navy had three times as many ships as the French. During the early stages of the conflict, France's revolutionary armies performed surprisingly well; any of their generals who failed to be successful was promptly executed.

In 1793 both Napoleon Bonaparte and Arthur Wellesley (the future Duke of Wellington) were 24 years old and lieutenant-colonels. Napoleon was very successful in Europe but was well aware that final victory would elude him if he could not add the occupation of England to his triumphs. That would be impossible if French ships could not break out of the blockade which the Royal Navy had imposed on the main French ports. Blockading a port was a difficult task in the time of sail, for adverse weather conditions could easily drive ships out of position. In the circumstances, Napoleon decided to wait for nature to break the blockade for him and in the meantime to confront the British in Egypt. Although the Suez Canal had not been created, Egypt was a valuable base and source of supplies. In 1798 Napoleon managed to slip out of Toulon, capture Malta and land at Alexandria. Nelson was prevented by a storm from intercepting him, but caught up and destroyed eleven out of the thirteen of Napoleon's ships at Aboukir Bay. Having overrun as much of Egypt as he needed, Napoleon now moved on to Syria where he besieged Acre. However, at this stage he felt that he was too far from the critical area of Europe and returned to France. He left a substantial army in Egypt but this was defeated by General Sir Ralph Abercrombie's army, which, although it was outnumbered, performed with great skill in the Battle of Alexandria. Here the Gloucesters fought back to back and thus earned the distinction of wearing a cap badge front and back.

After two devastating victories over the Austrians (at Marengo and Hohenlinden) in 1800, Napoleon tried to break the British naval grip on the Continent by coercing the Baltic countries into what he called the 'Armed Neutrality', which would have excluded British ships from the Baltic.

Gloucester. Relics of the 9th Gloucestershire Regiment are held in the regimental museum in Gloucester.

Copenhagen was the key to forcing a way into the Baltic and a fleet under Admiral Sir Hyde Parker was sent to bombard it into submission. For this purpose Parker used rockets which, although they were not a new invention, had been neglected for many years. The Danes were not seriously threatened by the rocket fire and Parker gave the order to retire. However, Nelson, who was one of the captains, put his telescope to his blind eye and said: 'I do not see a signal to retire.' He then forced the straits and made the Danes abandon their pro-French policy. It is from this event that the expression 'turning a blind eye' arises.

In 1802 a peace treaty was signed between Britain and France who were weary of war. However, neither felt that it was more than a temporary truce and war began again the following year, bringing Britain close to her hour of greatest peril.

In 1803 Napoleon began to muster an army of 120,000 men between Dunkirk and St Valery and he assembled thousands of flat-bottomed boats to transport it across the Channel. Landing drill was practised day after day and undoubtedly this was the biggest threat that England had faced since the Norman Conquest. The French waited for a gale to blow the Royal Navy out of its defensive positions or a fog to blind it. But the opportunity for invasion never came. For two-and-a-quarter years England waited, doubtless thanking God for the Channel. To counter this threat, an army of over half a million was raised, some of it virtually useless but most of it very able and determined. Meanwhile, French policies were creating enemies in the European territories that her armies had overrun. Napoleon, realising that an army of 120,000 cannot stand to arms for ever, decided to decoy away the British fleet by pretending to send the French fleet to attack the British possessions in the West Indies. Nelson was at first deceived, but recovered quickly and was soon able to confront the French fleet off Cadiz. Here, at the Battle of Trafalgar, on 25 October 1805, he destroyed his opponent ship by ship by brilliant naval tactics. Nelson was killed, but before he died he had crippled French seapower and thus saved Britain from the threat of invasion. 🔎

Napoleon was still equal to any challenge to his power in Europe, where he defeated Prussian and Russian attempts to confront him. In 1806, William Pitt died at the age of 46, exhausted by the strain of being prime minister for nearly half of his life in the most dangerous period the country had ever experienced. But the war went on.

> 🔎
> **Trafalgar Square, London.** Nelson's best-known monument is the column in Trafalgar Square, which is a familiar sight to millions.

CAMPAIGNS IN SPAIN AND PORTUGAL

Napoleon decided to draw off British military strength by opening a campaign in Spain and Portugal, which was to prove another of the most gruelling expeditions the British army has ever endured. There is a saying about Spain: 'It is a country where large armies starve and small armies get beaten.' The campaign opened with the French invasion of Spain and Portugal, at the end of which the French occupied Lisbon. Full of his own importance, Napoleon decided this would be an appropriate moment to make his brother King of Spain. However, he had not taken into account the Spanish character. Lethargic and easy-going though the Spaniards are, they are still a proud people. Once roused, they will fight like tigers and hang on until the bitter end. Napoleon soon discovered how wrong he had been about both Spain and Portugal, for both countries promptly rose in an insurrection. One of his corps commanders was defeated at Baylen and surrendered with 15,000 men; the other, commanded by General Junot, ran into the English troops which Wellesley had just landed in Portugal. (As mentioned, Wellesley would later become the Duke of Wellington, so for the sake of clarity we will refer to him by that title from now on.) Wellington won a tremendous victory at Vimiero on 21 August 1808, in a battle that lasted for three hours, but his commander-in-chief was a cautious general named Sir Hew Dalrymple who refused to allow Wellington to pursue the French. Wellington pointed out that his troops had plenty of rations and ammunition for the pursuit and begged to be allowed to continue. When Dalrymple refused, Wellington commented: 'There is nothing left for us to do, gentlemen, but to shoot red-legged partridges.' A great opportunity was lost but Dalrymple was relieved of his command and replaced by Sir John Moore, one of the pioneers of mobile infantry. The victory at Vimiero involved many famous infantry regiments. At the time they were still numbered but their later titles were: the Worcesters, the South Lancashires, the Highland Light Infantry, the Warwickshires, the Duke of Cornwall's Light Infantry, the Queen's, the Lancashire Fusiliers, the Oxford and Bucks, the Royal West Kents and the Rifle Brigade.

Glasgow. See the Museum of the Royal Highland Fusiliers in Glasgow for many interesting relics and exhibitions.

These two defeats infuriated Napoleon and he despatched to Spain an army of 250,000 men, most of whom were veterans who had brought him victories in Europe. He overcame the smaller Spanish armies without difficulty and then tried to advance to Cadiz and Lisbon. In this he was frustrated by Sir John Moore who, although his army was outnumbered by ten to one, swooped on Napoleon's communications. Napoleon then gave 100,000 of his troops the task of capturing Moore, but Moore very skilfully retreated, inflicting heavy losses on the French. Moore knew that with such a disparity in numbers he could never hope to defeat the French army, but he was determined to give it as much trouble as possible and to draw it away from its objectives.

While Moore was retreating towards Corunna, Napoleon heard that the Austrians, whom he had conquered three years earlier, had now raised another army; he therefore returned to deal with the new crisis, but left 200,000 troops in Spain. Moore continued his retreat to Corunna, where he planned to embark his troops. Before doing so, he turned sharply on the French and sent the vanguard reeling backwards, which enabled the British troops to embark and return to England for re-equipping. Moore himself was killed at the moment of victory. Moore's army re-equipped itself in England, set

Wellington relics at Apsley House, London.

sail again and landed at Lisbon where it was under Wellington's command. Although this army was only 20,000 strong, it defeated the French, first at Oporto and later at Talavera.

Meanwhile, Napoleon defeated the unfortunate Austrians once again and was able to send an additional 70,000 troops to Spain. A French general named Masséna was now given the task of driving the British into the sea. This seemed an easy task in view of the disparity in numbers, but Wellington had used the winter to construct a triple line of fortresses on the way to Lisbon. These were the famous 'Lines of Torres Vedras'. Wellington moved forward to meet Masséna's army well ahead of the Lines and then retreated slowly before him, destroying everything edible or habitable as he did so. Wellington's own army also suffered appalling hardships, but the French were in a worse state. Eventually, the British fell back into the Lines, which contained 600 guns and 30,000 British troops. Realising that he would not be able to break through unless he received reinforcements for his exhausted troops, Masséna camped in front of the Lines and waited. After four months he

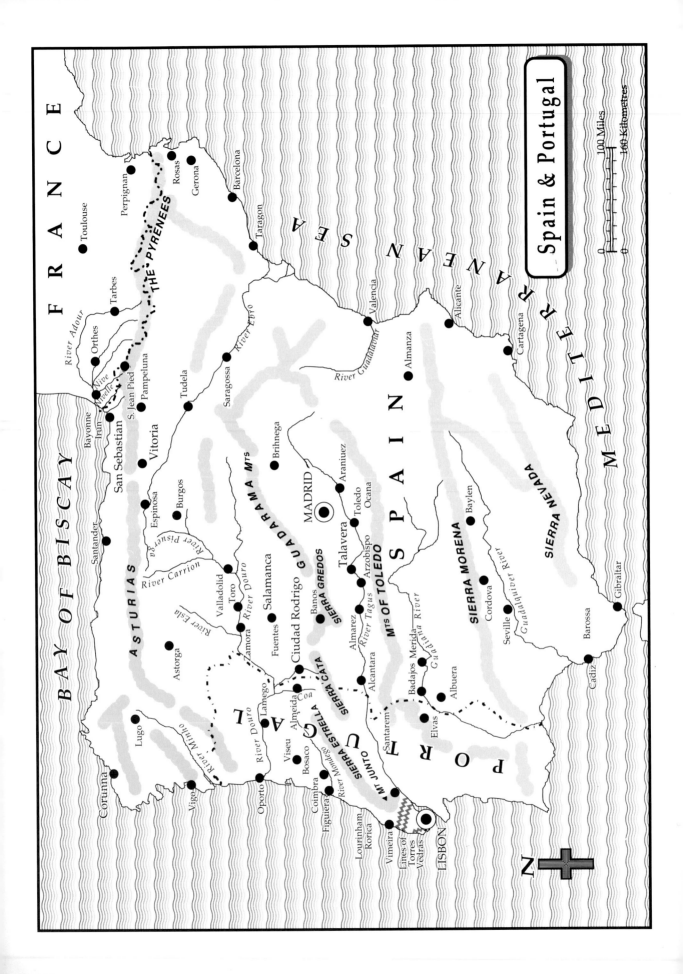

realised that the reinforcements were not coming; Napoleon, in fact, had other plans for them – an invasion of Russia.

Finally, in despair, Masséna began to withdraw his exhausted, hungry army back towards Spain. Wellington followed, perpetually harassing the French army and cutting off stragglers. In May 1811 Masséna turned, hoping to catch Wellington off-guard. His efforts failed and at the Battle of Fuentes d'Onoro he was defeated and was soon afterwards recalled, in disgrace, by Napoleon.

In the south another French army under Marshal Soult was defeated at Albuera. This was a narrow and expensive victory for the British; they lost 4,300 men out of 7,500. The battle hung in the balance until the Fusilier Brigade, which had only 1,500 men left, climbed a hill held by 7,000 French and by desperate close-quarter fighting forced them to surrender. Not surprisingly, Albuera Day (16 May) is an important date in the calendar of the Royal Regiment of Fusiliers.

It was also an important date for the Middlesex Regiment, who, although they are now absorbed into the Queen's, still have their own museum in Tottenham, London. During the battle the colonel of the Middlesex (then the 57th) was badly wounded and unable to stand. In spite of his weakness and pain, he constantly called out to his men: 'Die hard!' The regiment obeyed and from that date it was known as 'the Diehards'.

Albuera saw some spirited fights around the colours. When Ensign Thomas of the 60th Rifles was surrounded by Frenchmen, he refused to give up the colours and was killed. Ensign Walsh tried to carry the colours to the rear, but was wounded and captured. Lieutenant Latham then grabbed the colours before the enemy could take them. He was attacked and wounded in several places. A French hussar slashed at him, cutting off one side of his face and his nose, but he refused to give up the colours. Another hussar then cut off Latham's sword arm. As he was forced to the ground he managed to put the colours under him. Two squadrons of the British 4th Dragoon Guards then attacked the French hussars and Latham's body was trampled. At the end of the battle, when the colours were seen by a Fusilier sergeant, Latham was discovered to be still alive. He eventually recovered, although he was terribly disfigured.

DEFEAT AND EXILE

Napoleon should now have concentrated his army in Spain, which was conveniently close to France, and aimed at final victory there. Instead, he launched 600,000 men into Russia. He won a devastating victory at Borodino, although it cost him heavily in casualties, and he went on to capture Moscow. When he arrived in Moscow, the Russians, having learnt a lesson from Wellington's campaign in Spain, had laid waste and burnt the city. This left Napoleon with no alternative but to retreat. At the end of Napoleon's Russian campaign, only 60,000 out of the original 600,000 troops arrived back in France.

Although this was a major setback for Napoleon's grandiose plans, he refused to accept defeat and instead resolved on further conquests. The only source of replenishment for his battered European army was Spain and his army there was already staggering under the defeats that had been inflicted by Wellington, most recently at Ciudad Rodrigo and Badajoz. These were two fortresses on the frontier which Wellington had captured, although with heavy losses. The first of the battles took place in January 1812, in bitterly cold weather. Horses had nothing more to eat than chopped straw and the soldiers' rations were equally meagre. There were no tents. Wet clothes froze on the men at night as it was not possible to light fires to dry them. However, by tremendous efforts the fortress walls were scaled and the citadel was captured on 19 January. The battle was a special triumph for the Light Division, with the future Rifle Brigade, the 60th Rifles, and the Oxford and Bucks, but the Northumberland Fusiliers, the Middlesex and the Connaught Rangers also earned distinction that day. An equally fierce battle took place at Badajoz on the following 16 April, during which the British stormed the walls by using ladders as there was no time for a lengthy bombardment.

An even more remarkable battle took place on 22 July 1812 at Salamanca, where Marshal Marmont was commanding the main body of the French army. Unfortunately for the French general, in order to make an encircling movement, he had put his left wing in a vulnerable position, a fact which was quickly appreciated by Wellington. Instead of waiting for the French manoeuvre to be complete, the 'Iron Duke' fell on the French, who were widely dispersed, and cut them to pieces. One of Marmont's generals sub-

(Below left) The Duke of Wellington.

(Below right) Napoleon I.

sequently said that Wellington had beaten 40,000 men in forty minutes. From Salamanca, Wellington went on to capture Madrid which had been in enemy hands for four years.

Even after these crushing defeats the French could still rally a large army in Spain and for the rest of that year Wellington had the disadvantage of fighting superior numbers. However, in the spring of 1813 he attacked the French wherever he found them, culminating in the battle of Vittoria (21 June 1813) when he routed the French Marshal Jourdan and captured every cannon and waggon which he possessed. Among them was the carriage in which Napoleon's brother Joseph (whom Napoleon had made King of Spain) was trying to escape. Joseph abandoned his carriage, which was captured by the 14th Hussars, and escaped on a horse. The carriage held his silver chamber-pot, which was duly cleaned, burnished and subsequently used in the officers' mess, where it became a mazer for champagne. Champagne is a popular drink in the cavalry and drinking it out of a silver chamber-pot was probably easier than drinking it out of a lady's slipper. After this incident the 14th (later the 14th/20th) acquired the nickname of 'The Emperor's Chambermaids'. (The cavalry like to open champagne somewhat dramatically by removing the wire and foil, then running a sabre up the seam of the bottle. This neatly chops off the neck, complete with cork.)

Later that year Wellington captured the two frontier forts of St Sebastian and Pampeluna in 'The Battle of the Pyrénées'. He was now ready to invade France. After his disastrous Russian campaign

Joseph Bonaparte escaping at Vittoria in 1813, but leaving an important possession behind.

and failure in Spain, Napoleon was desperate. He continued to fight, but after Paris had fallen to the allies, his marshals forced him to abdicate. He tried to commit suicide but failed. The allies decided to spare his life and exiled him to the island of Elba, where he was allowed to rule its population of 10,000 Italian peasants. It was assumed – wrongly as it proved – that he would be incapable of disturbing the peace of Europe any more. The war had brought great distinction to Wellington, whose career appeared to have reached its peak when he had invaded France, captured Bordeaux and defeated his old rival, Marshal Soult, at the battle of Toulouse. Wellington never dreamt there was an even sterner test to come.

WAR WITH AMERICA

European countries now began to disarm, thinking that the twenty-two-year-long war was now over. However, England was not able to disarm herself as she was engaged in a little-known war with America, which lasted from 1812 to 1814. The cause of war was the Orders in Council which the British government had issued to prevent neutral ships from trading with French-controlled territories. They had been issued in retaliation for Napoleon's Berlin decrees, which prohibited neutrals from trading with Britain. In 1812, after five years of bickering, the American government declared war on Britain (although she had equally good reason to declare war on France). As Britain was already engaged in the Peninsular War, she had no spare troops to send overseas and the brunt of the fighting fell on the Canadian militia. The Americans tried to invade and subdue Canada on three occasions, but each time their forces were beaten back; on two occasions the invading army was forced to surrender by the gallant Canadians. However, Britain was by no means as successful at sea. The American frigates carried heavy armament which was skilfully handled; when they encountered British ships in single combat early in the war, they were more than a match for them. Although the American frigates were eventually hunted off the seas, it was not an occasion which the Royal Navy can remember with pride.

After Napoleon had been exiled, Britain was in a position to send more troops to America. Two of the expeditionary forces were ineptly led and failed, but the third, under General Ross, beat the American forces at Bladensburg and went on to burn Washington, which was then, as now, the capital of the USA. In the forefront of the Washington battle were the Black Watch, who, realising that the event had made them very unpopular, did not revisit the American capital for a ceremonial visit for a century and a half. However, when they did eventually return to Washington it was clear that all was forgiven if not forgotten. Peace was eventually signed on 28 December 1814, but resentment lingered on, in Britain because the Americans had attacked them when they were fully occupied with the French, and in America because the Americans felt that the burning of their capital was an insult they would not forget easily.

Rostrevor, County Down.
Ross' home town, Rostrevor, has a tall, grand obelisk erected in his memory.

THE BATTLE OF WATERLOO

At the beginning of 1815, when Britain was finally beginning to disarm, the alarming news arrived that Napoleon had escaped from Elba and had landed in France with 700 supporters. He had chosen just the right moment, for the restored King of France, Louis XVIII, was very unpopular, particularly with the army. When Napoleon called on the French army to remove Louis XVIII and to serve under him instead, the response was immediate and overwhelming. Seizing his opportunity, he announced that he was now a liberal rather than an autocrat, and that he had renounced war as an instrument of policy. His words were received more credulously in France than elsewhere and his former opponents promptly declared war on him once again.

As Napoleon's move had taken his opponents by surprise, he had the initial advantage of possessing 130,000 men, who were all in a state of combat-readiness. Against this the allies could only muster a force of Prussians, commanded by the ageing Marshal Blücher, and a mixed force under Wellington, which included British, German and Dutch. The English contingent numbered 30,000, the others 65,000. Napoleon decided that, if he struck quickly into Belgium, he could destroy Blücher's outnumbered force without difficulty and then he could turn and defeat Wellington's hastily assembled composite force. On 16 June he succeeded in the first part of this aim by defeating Blücher's Prussian army at Ligny, but he miscalculated in thinking that Blücher would then move the remnants of his battered force back to Prussia as quickly as possible. Instead the stalwart old Prussian set off to join Wellington, hoping to reach him in time to be of use.

Napoleon now turned his attention to Wellington's army. He learnt that Marshal Ney had encountered the British at Quatre Bras on the 16th and been repulsed, but was encouraged to hear that Wellington was preparing to give battle on a hillside at Mont St Jean, 12 miles (19km) north of Quatre Bras. Victory here would give him possession of Brussels and once more he would be master of Europe.

Prior to the Battle of Waterloo, Napoleon had 72,000 troops to confront Wellington's 67,000, but the latter figure included 20,000 Dutch and Belgian conscripts who had no military experience and no heart for a fight. Wellington deployed his army with the infantry in front and the cavalry partly on the wings and partly in reserve. In front of the English position were two farms, one at La Haye Sainte, which was garrisoned by the British, and one at Hougoumont, which was held largely by Hanoverians. Napoleon began by attacking Hougoumont but the Hanoverians beat him off. He then launched a four-column attack of 10,000 men on La Haye Sainte; it was held by 3,000 English under General Picton. As they approached Picton, the French were raked by British artillery and thrown into confusion by concentrated British fire. Just as they prepared themselves for a great effort to carry the position, Picton

Waterloo. Although only a few regiments have been mentioned here, visitors to regimental museums will see that many others have 'Waterloo' on their battle honours. Many will also display paintings and relics from the battle.

Dover Castle, Dover. Dover Castle has a layout of the Battle of Waterloo.

Stratfield Saye, Hampshire. The home of the Duke of Wellington (and home of the present Duke) is open to the public on weekdays in the summer, and weekends in winter. It contains a Wellington exhibition, his funeral carriage, and the grave of his charger Copenhagen.

Walmer Castle, Kent. This castle contains relics of Wellington, who lived there when he was Warden of the Cinque Ports. They include his coat, the original Wellington boots, and his narrow bed. One of Wellington's friends once remarked that he was surprised to see that he used a bed which was so narrow that there was no room to turn over. He received the reply, 'When one begins to turn over in bed it is time to turn out.' Most of the relics here were collected and presented to the castle by Wing Commander Lucas in 1966, but the room in which they are housed is preserved as it was in 1852, the year of Wellington's death. His adjoining bedroom is furnished with the original furniture, or restorations of it. It is an extremely simple room which includes a high desk, as he preferred to work standing up. His telescope, tea and coffee cups, and two bronze jugs are all there.

(Above) A piper in 1815.

(Below) A skirmisher. Cavalrymen used fast horses to ride up to or even through the French troops to gather information about their strength and quality; usually the French could not catch them.

The Battle of Quatre Bras.

The Royal Dragoons capture a French eagle.

(Opposite page) Quatre Bras. High rank was no protection in close-quarter fighting.

Cavalry charges broke up formations at Waterloo and added to the confusion.

The 28th (1st Gloucester) Regiment receiving French cavalry at Waterloo and checking them on 18 June 1915.

(Opposite, above) During the early stages of the Napoleonic Wars fortresses were built along the south coast of England, many being named 'Martello' towers. The Wish Tower was one of them, and is now a museum.

(Opposite, below) The Redoubt Fortress, Eastbourne.

ordered his infantry to charge. He himself was killed in the opening moments of the advance, which came to a halt when it ran into a solid body of French troops. At that point the Uxbridge's Union Brigade, the Royal Dragoons, the Scots Greys and the Inniskillings were brought into the attack and, after moving through the British infantry, crashed into the French lines. The force and spirit of this charge threw the French into disarray, which was followed by despair. Finally, the French fled from the battlefield leaving many dead and 3,000 prisoners; several French eagles were also captured. Two paintings of the charge of the Scots Greys, one by Lady Butler, the other by Stanley Berkeley, give a vivid impression of the vigour and thrust of the battle.

Napoleon was not a leader who worried about losses on his own side and now decided to concentrate the remainder of his forces on the British centre. The British were drawn up in squares, which the French pounded with artillery and charged relentlessly for five hours. The Dutch and Belgian conscripts were worn down by the pounding and retreated, but the British held their positions. In spite of the desperate nature of the fighting, many officers displayed chivalrous conduct when these old enemies met for the last time. There were also extraordinary examples of fortitude. Lord Somerset, later to be Earl Raglan, commander-in-chief at the Crimean

National Army Museum, Chelsea, London. You will find many mementoes relating to the Battle of Waterloo here, including a number of excellent paintings.

Royal Artillery Museum, Woolwich, London. Here are examples of the larger guns used in Wellington's campaigns.

Royal Military Academy, Sandhurst, Camberley. In front of the old building are cannons captured from the French at Waterloo. Sandhurst also contains good paintings of the battle and the personalities of Waterloo – there is one of Copenhagen, the Duke's charger, and another of Norman Ramsey, a young Royal Horse Artillery officer who was killed in the battle.

Wellington College, Berkshire. This public school was founded in memory of the Duke, and has special school places for boys whose fathers were killed or died on active service.

War, was so badly wounded in the right arm that it had to be amputated (without anaesthetic). Lord Uxbridge, who was riding close to Wellington, had his leg taken off by a passing cannonball. 'By God, I've lost my leg,' he said. The duke put down his telescope for a moment and looked at him. 'By God, you have, sir,' he answered. The remains of the leg were then amputated with a crude saw which is now in the National Army Museum, London. Uxbridge survived, and, having been fitted with a wooden replacement, became known as 'Old Peg-leg'.

While the battle still hung in the balance, the remains of Blücher's army, which Napoleon had defeated at Ligny, now advanced. They were a welcome sight to the British, whose losses had been heavy. However, Napoleon was not finished and sent in 5,000 men of his 'Old Guard', veterans who had won battles all over Europe. These drove forward, suffering heavy casualties, until they almost reached the crest of the hill at the centre of the British position. At that point they were held by British infantry, who refused to withdraw from their position. The French then faltered, broke and began to fall back. At that moment Wellington ordered another cavalry charge in which the 16th Dragoons (now Lancers) and the 11th Light Dragoons (now the Hussars) and the 18th Hussars (now part of the 13th/18th) took the lead. It decided the battle.

Casualties on both sides were high. Wellington's army is said to have lost 22,000 men, but the French lost about 40,000. It was the end for Napoleon. He tried to retreat to Paris and raise another army but failed. Eventually he surrendered to the British, who sent him to St Helena, where he lived for a further six years.

The importance of Wellington's victory cannot be over-emphasised. Napoleon was a military genius, but he was also a fanatical megalomaniac tyrant; had he not been defeated Europe would have suffered for many years under the autocratic regimes he established. Of the final battle, the result of which hung in the balance all day and which was not decided until late evening, Wellington said: 'It has been a damned serious business – Blücher and I have lost 30,000 men. It has been a damned nice [close] thing – the nearest run thing you ever saw in your life. By God I don't think it would have done if I had not been there.' He also said: 'Hard pounding this, gentlemen, let's see who will pound longest,' but he denied that he ever said, 'The Battle of Waterloo was won on the playing fields of Eton.'

Waterloo was one of the most important battles in history and regiments through Britain are very proud to have taken part in it. Among the many places which commemorate the battle is Waterloo Station. The expression 'to meet one's Waterloo', signifying to encounter a superior force, is now a common English idiom. Numerous public houses are entitled 'The Iron Duke', and bear Wellington's picture on their signboards.

CHAPTER 6

DISTANT BATTLES

Whilst the life and death struggle with Napoleon was being waged in Europe, Britain was involved in other campaigns which took place mainly in India and the surrounding countries; these campaigns are commemorated on the battle honours of various regiments. While he was in India, before being brought back to Europe to campaign against Napoleon's forces, Wellington fought against and defeated the Mahrattas, who had formerly dominated northern India. The collapse of the Mahratta empire led to large groups of men terrorising central India, who were not subdued until 1825.

During this period Britain was also involved in war with the Gurkhas of Nepal, who had gradually been expanding their territory into India. The battle with the Gurkhas ran from 1813 to 1816 and was only won by the most desperate fighting. However, during the course of the conflict, both sides developed a liking and admiration for each other, which resulted in Gurkha troops becoming an integral part of the British army. Since that time Gurkhas have fought side by side with British troops in campaigns and wars all over the world.

Gurkha Museum, Winchester. The Gurkha Museum contains a pictorial history of the Gurkha part in campaigns throughout the world, as well as a magnificent collection of kukris, medals and badges. The Gurkhas have won 26 Victoria Crosses and here you will find descriptions of the actions in which they were won.

During the nineteenth century Britain was involved in a series of wars which took her armies into remote and extremely difficult territory. From the fifteenth century onward European explorers, initially seamen, had made contact with distant countries and tried usually successfully, to build up trading partnerships. In the course of several hundred years, Portugal, Spain, France and Britain had extended these 'partnerships', usually selling manufactured goods, such as clothing or metalwork in exchange for raw materials or even gold, silver, or precious stones. Some of the countries, though not all, were very underdeveloped. At the time, few people questioned the right of European countries to populate, annex, make treaties with, or colonise overseas territories. Most of these were thought to benefit from the imposition of European law and commercial development. Sometimes one European country would fight another for possession of territory of commercial or strategic importance, as the French and British did in India and Canada. On occasions a European settlement, made by treaty with the local chiefs, would be attacked by a hostile neighbour; this often resulted in a minor war, and the annexation of the former hostile neighbour's territory.

In the twentieth century imperialism of any form has been condemned by the developed world, although the USSR still controls some of the countries her armies occupied in Eastern Europe at the end of World War II. In addition, many parts of the world are 'colonised' by economic imperialism.

The Ross obelisk at Rostrevor, County Down, Northern Ireland, commemorates the General Ross who captured and burnt Washington in 1814, after the Americans allied themselves with the French in 1812. When peace was signed in December 1814, both sides gave up their conquests.

(Opposite page) Hilt of an officer's broadsword of the 79th (Cameron) Highlanders *circa* 1820.

Simulated battle scene in Gurkha Museum, Winchester.

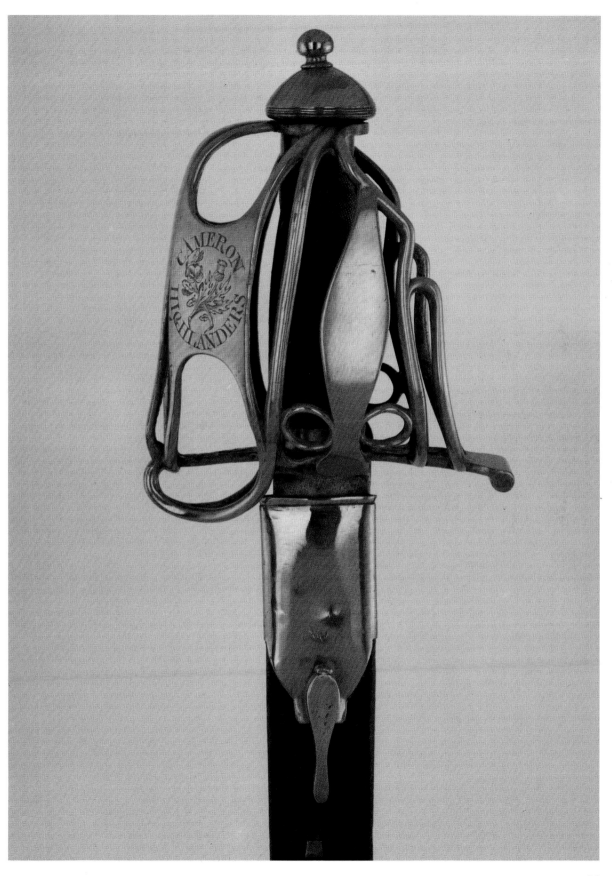

Until the twentieth century, therefore, there have been numerous small wars undertaken with the aim of economic or military expansion. Most former colonies have since been granted independence, but unfortunately this has sometimes meant a short-lived period of democracy, followed by one-party rule, tyranny, or even anarchy and continuous civil war. The soldiers who fought in the gruelling campaigns of the nineteenth century knew little or nothing of the reasons why they were fighting. Their regiments had been sent to win campaigns and battles. A century later someone would have explained why they were there and what the political objective was, but these earlier soldiers simply knew they had been given a task which, when successfully performed, would be commemorated on the battle honours of their regiment.

One of these earlier campaigns was the Burma War of 1824 to 1826 which occurred when the Burmese invaded India and claimed Eastern Bengal. Lower Burma, as later British troops learned in World War II, is swampy, covered with thick jungle and very unhealthy; before the days of modern antiseptics and anti-malarial tablets, it could be devastating to armies. Even in 1942 some units were reduced to half their numbers by tropical diseases.

A group of Scottish soldiers in 1846.

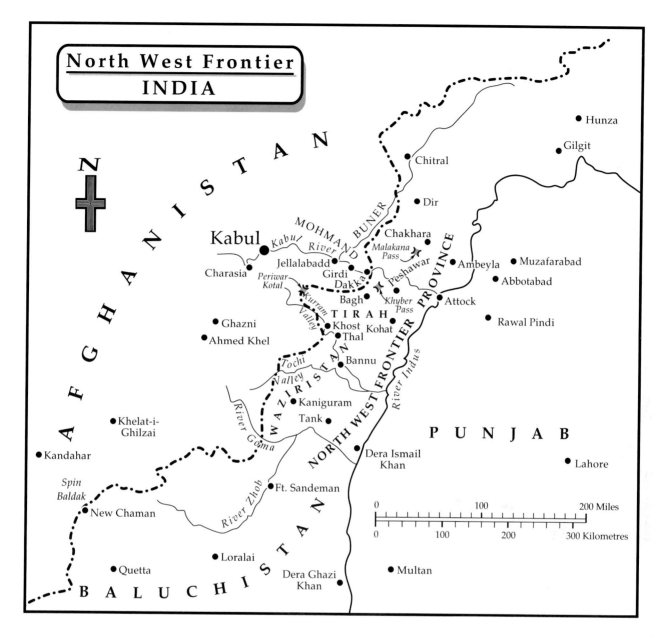

North West Frontier
INDIA

The local troops which the British encountered in Burma in the late 1820s were well armed, possessed excellent artillery and were very skilled at constructing stockades and other defensive positions. However, in spite of heavy casualties from disease and battle, the British eventually secured victory.

Another, less successful, war took place in Afghanistan some ten years later (1838 to 1842). In the 1830s the British government became extremely suspicious that the Russians were making diplomatic and other inroads into Afghanistan with a view to using that country as a springboard for the invasion of India and Persia. Afghanistan was in a state of chronic internal strife and it seemed to the British government that Britain should support one of the claimants to the throne who, presumably, would then look upon

(Above) An officer's dirk of the 74th
(Highland) Regiment in 1846.

Mid-nineteenth century cavalry; they
looked splendid even when
dismounted.

Britain with gratitude. Although the British candidate was duly installed, the whole of Afghanistan then rose against the British. It was thereupon decided that it was futile for British forces to stay in a country which was so hostile to outside intervention and that they should be withdrawn. The British army was promised a safe passage to retreat, but this was not honoured: one garrison was massacred after it had surrendered and another narrowly escaped a similar fate. A relieving force was sent to retake Jalalabad, which was besieged, and soon Kabul too was captured. The British then withdrew.

Heliographs were widely used for signalling in India, where they had a range up to 90 miles.

Thirty-six years later Britain was involved in another Afghan war for the same reasons; that the Russians were making incursions into Afghanistan and were being encouraged by the king, Sher Ali. Britain's military tactics in the second Afghan war were more professional and involved nearly 60,000 men. The initial conquest, led by General 'Bobs' Roberts, was successful and a peace treaty was signed in 1879. The Afghans agreed to receive a British mission in Kabul and not to embark on foreign policy ventures which might prejudice the safety of India – that is, they would not make any dangerous treaties with the Russians. The main British force then withdrew, leaving only small garrisons. This policy once again proved disastrous for there were insurgent groups which could easily muster enough armed men to overwhelm the small garrisons. One of the most dramatic events in the campaign occurred when the Berkshire Regiment (now combined with the Wiltshire Regiment to form the Duke of Edinburgh's Royal Regiment) fought a desperate action at Maiwand, where the British were outnumbered by ten to one. Some of them managed to reach the safety of the town of Kandahar, but the Berkshires (the 66th, as they still were then) were reduced to eleven men, who finally charged and were all killed.

The war in Afghanistan was eventually brought to a successful conclusion by Roberts. The Emir (king) Abdur Rahman was then left in undisputed possession of the territory. He proved to be a staunch friend of the British for over forty years.

Other wars which took place in the first half of the nineteenth century included the conquest of Sind (a huge province which is now part of Pakistan) and desperate battles against the Sikhs, who lived in the territory between the Indus and Sutlej. The Sikhs were a deeply religious, warrior people and it was hoped that the British could make a treaty with them and settle down to a co-operative existence. However, the Sikhs thought otherwise and marched into Sind with 50,000 men and 100 guns in December 1845. The ensuing war lasted for four years, and involved many fierce battles and several British setbacks. Subsequently, many Sikh regiments earned distinction fighting in the Indian army.

Although by the mid-nineteenth century the British were settled in Lower Burma and trade was flourishing, the local dignitaries resented the presence of foreigners and imposed a number of regulations which made trade almost impossible. The merchants thereupon appealed to the Governor-General of India to intervene on their behalf. A diplomatic team was sent to Rangoon to negotiate, but the frigate on which it was being carried was fired on before it could come to port. From then on the position deteriorated rapidly. A small army, numbering less than 6,000, was despatched, and after a series of guerrilla-style battles, this forced the Burmese court to agree that British traders should not be molested. The promise was not kept and in December 1852 Lower Burma was formally annexed to India after a difficult campaign.

Cathedral Close, Salisbury. The Duke of Edinburgh's Royal Regimental Museum can be found here.

Forbury Park, Reading. The achievement of the Berkshires is commemorated by the 'Maiwand Lion', a stone memorial in Forbury Park, Reading, as well as in the regimental museum at Salisbury.

All the above campaigns were lengthy, difficult and extremely arduous. The fact that they were won at all was a great tribute, not merely to the infantry, but also to the supply services and medical corps. In Burma, the centre of government was 500 miles (805km) from the coast; reaching it involved building bridges, transporting food and ammunition, and battling with various fatal diseases, such as cholera, whose causes were not yet understood. In India, opponents such as the Sikhs had modern arms which they had obtained from the French. Some countries voluntarily became part of the British Empire because they realised that that course was their only hope of survival among more powerful neighbours. When Britain took over a territory it introduced much needed reforms, such as the establishment of law and order, roads, bridges, irrigation, communications – for example, railways – and the provision of health services.

The Indian Mutiny of 1857 was therefore a great shock as well as a major disaster. It was a rising of the Sepoys (privates) of the Bengal army and did not spread to the Madras and Bombay areas. The Sikhs and the Mahrattas also stayed loyal to the British. The mutiny began with bloody massacres and lasted for nearly two years. Undoubtedly it was stirred up by political agitators, but it also indicated that more reforms were needed and that the British should not become complacent in the territories they were governing.

The fact that the Indian Mutiny came as a complete surprise to the British gave the rebels an enormous initial advantage. They had been trained by the British and had modern arms and equipment. Furthermore, as they greatly outnumbered the British, they were able to seize vital strategic points. Winning these back and defeating the rebels in fiercely fought battles brought many honours to British regiments. The rebels seized Delhi – which contained a huge supply of arms – massacred the Europeans in the city and used it as their headquarters. Lucknow, Allahabad, Benares and Cawnpore were all besieged. After a three-week siege, the 800 inhabitants of Cawnpore surrendered, having been promised their lives. They were promptly massacred; their number included many women and children. However, one after the other, these towns were all relieved by British and loyal Indian forces after bitter fighting and heavy casualties.

Among the regiments which distinguished themselves at Delhi were the 8th (now King's), the 75th (the Gordon Highlanders), the Gurkhas, the KRRC, the Oxfordshire Light Infantry (now part of the Royal Green Jackets), and the Royal Engineers, who blew up the gates. At Lucknow the 32nd (Duke of Cornwall's Light Infantry) were to the fore, as were the 64th (Staffordshire Regiment), the Northumberland Fusiliers (now part of the Royal Regiment of Fusiliers), the Royal Artillery, the Seaforth Highlanders (now joined with the Camerons to become the Queen's Own Highlanders), the Argyll and Sutherland Highlanders, the 53rd (1st Shropshire Light Infantry) and the Royal Engineers.

Sandhurst, Camberley. After the Indian Mutiny many changes were made in the organisation and administration of India. Many successes by the British army in later campaigns were won fighting side by side with comrades of the Indian army. Many museums contain relics of regimental service in India, and Sandhurst has a magnificent Indian Army Memorial Room which contains interesting exhibits and paintings.

CHAPTER 7

THE CRIMEAN WAR

The Crimean War, which lasted from 1854 to 1856, was fought on the Crimean Peninsula, many thousands of miles from Britain, on territory which was infested with deadly diseases such as cholera, and where the temperature is extremely hot in summer and bitterly cold in winter. The war caught the British government totally unprepared and this lack of foresight caused extreme suffering among the British troops.

The nominal cause of the Crimean War was an argument between Russia and France over which country should hold the key to the Church of the Holy Sepulchre in Jerusalem. The Sultan of Turkey, in whose empire Jerusalem then lay, was a weak ruler, and Turkey itself had long since declined from being the powerful imperialist nation which had once ruled with an iron hand through the Middle East as far as Egypt. The Turkish empire was now in such turmoil that the Czar of Russia, the belligerent Nicholas I, described it as 'the sick man of Europe', and he tried to persuade England and France that it should be divided up before it fell into complete anarchy. The Czar's attitude was diagnosed by France and Britain as one of cynical disregard for Turkey's real problems and as an excuse for Russia to seize large quantities of territory from a weak neighbour. The Crimean Peninsula, where Russia had a naval base at Sebastopol, was in the Black Sea, and the entrance to the Black Sea from the Mediterranean was through the Dardanelles and the Bosphorus, which were both in Turkish territory. If Russia gained control of Constantinople (now Istanbul) and thus of Turkey, Russian warships would have free access to the Mediterranean and therefore would be able to interfere with British and French trade routes and strategic communications.

When the Sultan of Turkey settled the dispute concerning the key of the Church of the Holy Sepulchre by deciding in favour of the Roman Catholics (who were supported by France), the Czar demanded impossible concessions from the Sultan and, when these were rejected, declared war. Russian troops crossed the Danube and the Russian navy destroyed the Turkish fleet. Britain and France thereupon demanded that Russian troops should be withdrawn from Turkish soil. The request was ignored and war was declared in March 1854.

Britain then sent an army of 28,000 to capture Sebastopol by landing 33 miles (53km) north of it and approaching it by land – it would have been impossible to capture the port directly from the sea.

After thirty-one years of neglect, the British army was totally untrained for a campaign of this type. The soldier had a uniform which was barely adequate for the temperate climate of Britain, it restricted movement, was conspicuous to enemy marksmen and was usually full of lice. His food was a stew of stringy beef and potatoes; the only variation in his diet was salt pork. In fact, the British soldier was probably less well off than his counterpart a hundred years earlier. By the time that the army landed in the Crimea its numbers had been drastically reduced by cholera, a water-borne infection which causes a violent diarrhoea. The weight of the unfortunate victim is usually reduced by half, or even more, before he dies a few hours later. It is not uncommon for a man to be walking about, apparently healthy, suddenly to be taken ill and be dead three hours later. Nowadays his life would be saved by inoculation and saline drips, but in the mid-nineteenth century there was no treatment: he either died or, miraculously, got better.

Very few who contracted the disease in the Crimea survived. Even when the troops had landed and begun their march to the heights of Alma, north of Sebastopol, they were still dying in significant numbers. Then, in the peculiar course which cholera follows, it suddenly ceased to claim any more victims. The extraordinary behaviour of cholera caused many people to believe it must be carried by an invisible vapour, called a 'cholera cloud'.

All the soldiers in the armies in the Crimea – British, French and Russian – were now using Minié rifles. These were muzzle-loaders which fired a conical bullet, reasonably accurately, up to 1,000yd (914m).

The first battle in the Crimea took place on the slopes rising from the Alma river after the British troops had marched over a treeless plain under a blazing sun without water. The attack began with artillery exchanges and then saw the Grenadiers, the Scots and the Coldstream Guards begin to climb the hill, still maintaining perfect parade-ground formation. Lord Raglan, Commander-in-Chief (who, as Lord Somerset, had his arm amputated at Waterloo) became separated from the army and lost touch with it. Consequently, the divisional commanders issued what orders they could, although these were often conflicting and caused confusion.

However, the Scots Fusiliers ✍ and the Welch Fusiliers were also pressing forward and soon the Russians were hit by the 42nd (the Black Watch), the 93rd (the Sutherlands) and the 79th (the Camerons). Meantime, 3,000 Russian cavalry stood idly by; clearly there was a lack of co-ordinated orders in the Russian army as well as in the British. Finally, when the Russians were in full retreat, Raglan refused to allow the British cavalry to pursue them. Cavalry can perform a very useful function in turning a victory into a complete rout and will be very frustrated when it is not allowed to to so. However, it has been suggested that although the cavalry might have succeeded in destroying the Russian army completely, it might equally have lost control of the situation and taken heavy casualties in unknown country against desperate opposition.

If the victory at the Alma had been followed by a full-scale assault on Sebastopol, the war would probably have been won before the

✍
Museum of the Royal Highland Fusiliers, Glasgow. This museum contains many relics of the Royal Highland Fusiliers.

end of the year. However, Raglan decided that such bold tactics were too risky and instead decided to march all the way around Sebastopol on the landward side and then to assault it from the south. His decision meant that he had lost the advantage he had gained at Alma and had also wasted time which enabled the Russians to strengthen the city's fortifications.

To the south of Sebastopol lay the port of Balaclava, which now became the British supply base. It was not a good port but it was vital to British success; it lay 6 miles (10km) from the position where the bulk of the British army was now deployed. French and Turkish troops were also on the peninsula, but so far neither had taken much part in the campaign. Since losing the battle on the Alma the Russians had made rapid efforts to assemble a new and large army with which to destroy the allied forces which had now assembled to the south of Sebastopol. On 25 October, an army of some 40,000 Russians advanced towards the British, with the intention of capturing Balaclava.

Part of the Russian cavalry advanced towards Balaclava itself, which was defended by 550 men belonging to the 93rd (which, with the 91st, later became the Argyll and Sutherland Highlanders). The 93rd were commanded by a doughty old warrior named Sir Colin Campbell. He told them to hold their fire until he gave the order and added: 'Remember, there is no retreat. You must die where you stand.' His words were greeted with a cheer. As the Russians advanced he gave the order for the first volley. The Russians checked but continued to advance. Campbell then gave the second order. This time the Russian commander ordered his men to wheel to the left, with a view to outflanking the 93rd. As they began their move, the 93rd fired their third volley; the Russians retreated.

It was essentially a test of nerve. The Russians probably realised that if they had continued to advance they would have broken through but would have experienced heavy casualties; their commander probably did not wish to sacrifice his troops in a battle which could be won later at less cost. He may have assumed that there were other equally resolute British troops immediately behind the 93rd. Whatever the thinking, there is no doubt that Colin Campbell's 93rd saved the vital port of Balaclava, with what came to be called 'The Thin Red Line'.

THE CHARGE OF THE HEAVY AND LIGHT BRIGADES

The Charge of the Heavy Brigade and the Charge of the Light Brigade both occurred on 25 October 1854. Neither of these was a brigade in the modern sense of the term (which means a unit of some 3,000 troops). The Heavy Brigade numbered 800 men and consisted of eight squadrons. 'Heavy' cavalry were mounted on larger horses and were usually brought into the battle when crushing charges were needed; however, they could also perform the other functions of cavalry, such as reconnaissance, skirmishing, harassing,

(Opposite page) The Heavy Brigade begin their charge.

The Heavy Brigade defeating the
Russian cavalry.

etc, when required. The Heavy Brigade was commanded by General
the Hon J. Scarlett and comprised two squadrons of the Inniskilling
Dragoons, two of the Scots Greys, two of the 5th Dragoon Guards,
one of the Royal Dragoons and one of the 4th Dragoon Guards.

Scarlett was waiting for orders from Raglan when suddenly he
saw the main body of the Russian cavalry coming over the nearby
Causeway Heights. It appeared to be a large army – in fact, there
were 3,000 troops. Observing the Russians moving very slowly,

obviously uncertain of their objective, Scarlett wheeled his left column of 300 sabres (swordsmen on horseback) and led the charge himself. He headed for the centre of the Russians and, when he arrived, engaged in combat with the best of them, although he was 61 years old. He himself received five wounds and his ADC fourteen. Meanwhile, the British artillery had found the range of the rear of the Russian cavalry and were beginning to shell it with devastating results. The Russians, believing that this was probably the first of a succession of charges, gave ground and finally fled. It was a magnificent performance by the 'Heavies', but unfortunately was overlooked after the battle when all attention was focused on the heroic, but virtually useless, charge by the Light Brigade.

The commander of the Cavalry Division, which consisted of the Heavy and Light Brigades, was Lord Lucan; he was related to the Earl of Cardigan, commander of the Light Brigade, but the two men despised and detested each other. Cardigan, who was extremely brave, was even more obstinate and stupid than Lord Lucan. When the Russian cavalry were in retreat following the Heavy Brigade's charge, Cardigan should have harassed them with his own brigade, but in fact he did nothing. Raglan, observing this inactivity, was not pleased, particularly when he saw that the Russians, who were obviously delighted at not being pursued, were beginning to haul away some Turkish guns from positions which had been overrun earlier. He therefore sent a message to Lucan, instructing him to 'prevent the enemy from carrying off the guns'. From where he was, Lucan could not see the Russians carrying off any guns at all and assumed that Raglan meant the guns which the Russians were manning at the far end of the valley. The message was delivered to Cardigan by a conceited captain, named Nolan, whom he disliked. Although Cardigan thought the order was insane he felt that he had no option but to obey it. He therefore set off with 673 men to ride 1½

The Charge of the Light Brigade.

miles (2km) up the valley to the Russian battery. It was a heroic occasion and entirely futile; it lasted twenty minutes. The general impression, created by poems and misleading accounts, was that 600 men rode up the valley and were all killed. In fact, of the 673 men only 113 were killed and 134 seriously wounded. The brigade never wavered as it approached the Russian guns, although it was also harassed by fire from the sides; having reached the guns and sabred the gunners, it could do nothing but turn around and ride out again. Cavalry cannot stop, dig in and hold ground which has been won – that is a task for the infantry.

Cardigan survived, unwounded, and after the charge returned to his yacht, which was anchored off the coast. There he had dinner, drank his customary bottle of champagne and went to bed. He does not appear to have concerned himself with the welfare of the men who had followed him into almost certain death. However, in this he was unusual, because most officers would have stayed on the battle-field to help the wounded.

Subsequent opinion has criticised the charge of the Light Brigade as being no more than a stupid mistake which led to the death of many brave men and the loss of some 600 horses. As far as imme-diate results were concerned, it was certainly pointless, but as an example of dauntless courage it impressed the Russians so much that they never set their cavalry against the British for the remain-der of the war. No doubt the performance of the Heavy Brigade also influenced their decision.

The Light Brigade consisted of the 13th Light Dragoons (later to be renamed the 13th Hussars and later still to be amalgamated with the 18th to form the 13th/18th Hussars), the 4th Hussars and the 8th Hussars (who would later be amalgamated to form the Queen's Royal Irish Hussars), and the 17th Lancers who would later join the 21st Lancers to form the 17th/21st Lancers.

Cannon Hall Museum, Barnsley, South Yorkshire. The 13th/18th Hussars have a part of the Cannon Hall Museum, from which area many of the regiment's soldiers have come.

Belvoir Castle, Lincolnshire. This is the home of the 17th/21st Lancers museum.

Edinburgh Castle, Edinburgh. The Scots Greys, who are now amalgamated with the 3rd Carabiniers to form the Royal Scots Dragoon Guards, have their museum in Edinburgh Castle.

THE SIEGE OF SEBASTOPOL

Although the Russians had made no progress in their attempts to push the British troops off the Crimean Peninsula, they were by no means disheartened and began to prepare a huge force for the next round in the conflict. The army they now assembled for this pur-pose totalled 55,000 and had 220 large guns; with these numbers it was likely to outnumber its opponents by two to one. However, they launched their initial attack with 35,000 troops. By confronting the British on the Inkerman ridge they believed that they could elimi-nate them before they moved on to destroy the French.

Although the British were well aware that an attack was impend-ing, it took them by surprise when it began at 5am in fog and dark-ness on 5 November 1854. As the Russians came into the British forward areas, the Russian guns began to shell the rear of the posi-tion. The Russians hoped that the British would immediately fall back and be destroyed by the Russian gunfire. However, the British troops who prepared for battle hurriedly in the general alarm had

different ideas and, instead of retreating, carried the war to the enemy. In the fog and the darkness it was impossible for the British officers to gauge the situation or to give practical orders; instead they had to let the soldiers fight without direction. Consequently, Inkerman became known as 'the soldiers' battle'. In the front of the fighting was the redoubtable 88th (Connaught Rangers) and close by were the 77th (the Middlesex Regiment), the 'Die Hards'.

Although the Russians had made some gains at the beginning of the attack they soon lost them to the Berkshires and the Welch Regiment. The Guards Regiments fought ding-dong battles, as did many others. Among those who covered themselves with glory that foggy morning were the Lancashire Fusiliers, the Sherwood Foresters, the Royal Scots Fusiliers and the Border Regiment. When dawn came, the fighting became even more intense.

At the end of the day, the casualty lists were long but so were the recommendations for bravery. As a result of this hard-fought conflict the Russians decided against taking on the allies in a set-piece battle and instead concentrated on harassing them as they laid siege to Sebastopol. The war then settled down into the trench warfare we should see in 1914–1918 and again in the Korean War of 1950–1953. During the Crimean winter both sides suffered greatly from the cold, especially the British. The scandal of inadequate clothing and food shocked the nation, but an even greater outcry came when the conditions of the hospitals were revealed. Florence Nightingale, a nurse of iron will and considerable influence, went to the Crimea and supervised the medical care of the sick and wounded; conditions improved, but not before many soldiers had suffered and died unnecessarily.

The war ended after desperate assaults on two fortresses at Sebastopol known as the Redan and the Malakoff. It was not the end of Russian ambitions, but it prevented their warships from entering the Mediterranean for over a hundred years.

Memorials to the Crimean War are not merely to be found in museums and regimental chapels. The battles are commemorated in the names of streets and houses (Alma Terrace, Sebastopol Place), and, of course, in the names of public houses. The inn sign 'The Hero of Inkerman' was once very popular, but today few people have heard of Inkerman, and the word has been quietly dropped.

Perhaps the finest memorial to the Crimean War was the reforms made in the army and in the nursing services, both of which were the result of the determination of Florence Nightingale. Florence Nightingale was a well-educated woman from a wealthy family. When she decided to become a nurse her family were horrified and did everything to prevent it, as nurses were held in very low esteem at that time. However, she was determined to follow her chosen profession and was eventually sent out to the Crimea. When she arrived at the Barracks Hospital at Scutari the doctors refused to allow her three nurses into the wards. Gradually she wore down the

(Opposite) The Guards at Inkerman.

Lower Regent Street, London. Here you will find an impressive memorial to the Crimean War.

National Army Museum, Chelsea, London. This Museum contains many Crimean relics and paintings, as well as an enormous stuffed cat. Its history is that it was found nearly starving in Sebastopol when the British troops entered the city. British soldiers have a long tradition of befriending stray animals, and this one lived a contented life with the troops until it died.

The Crimea. When visiting country houses and castles you will sometimes encounter trees which were brought back from the Crimea after the war and planted in British soil.

A scene in the Florence Nightingale Museum, St Thomas's Hospital, London.

The Assault on the Redan,
Sebastopol 1855.

Claydon House, near
Winslow, Buckinghamshire.
Here you will find a Florence
Nightingale Museum which
contains her bedroom and
sitting-room. It is now owned
by the National Trust and is
open to the public. There is
also a Florence Nightingale
Museum at St Thomas'
Hospital, London.

opposition, reorganised the service and transformed the hospitals
into places where men were healed rather than died. She herself
made the rounds of the wards every night, tending and talking to
the wounded. This gave her the nickname 'The Lady with the Lamp'
and was the subject of many pictures.

On her return to Britain she refused all receptions and celebra-
tions of her work, and set about reforming both army health and the
nursing profession. She set up the Nightingale School for Nurses at
St Thomas's Hospital, London: it was the first of its kind in the
world. Her reforms were vast and far-reaching. She had indicated
that she did not wish to be buried in Westminster Abbey and she
now lies in the family grave at East Wellow, Hampshire.

CHAPTER 8

AFRICAN AND OTHER WARS

In order to understand why the British army became involved in a series of gruelling battles in Africa it is necessary to give a brief account of the background to the European settlement in that troubled country, first southern Africa then the arduous campaigns which were fought further north.

CAMPAIGNS IN SOUTH AFRICA

South Africa had been discovered by Portuguese navigators in the late fifteenth century. Vasco da Gama landed at Natal on Christmas Day (Dies Natalis in Latin) 1497 and named the district accordingly. British ships soon followed and one British seaman claimed the Cape of Good Hope for Britain in 1620, only to find that the British government firmly refused to accept it.

In 1632 the Dutch East India Company, which had a steady trade with what is now Indonesia, decided that the Cape was a convenient halfway point and established a settlement there. They were soon in conflict with the Hottentot inhabitants, whom they often enslaved. During the Napoleonic Wars when the French overran Holland, Britain had to send ships to the Cape to prevent Napoleon's fleet from occupying it. In the peace treaty at the end of the wars it was decided that Britain should keep the Cape but pay the Dutch government £6 million in compensation. This suited everyone except the Dutch settlers around Capetown who moved north where they were in conflict with another native race, the Bantu.

The Bantu had no stronger claim to South Africa than the Dutch or the British because they had come from central Africa in tribal migrations and were already divided into groups which were hostile to each other. The Bantu would raid Boer farms and steal their cattle, while the Boers would capture the Bantu and Hottentots and enslave them. The Boer farmers at first looked to the British to drive the Bantu back, but, when the British were unwilling to send soldiers to engage in such an expensive and complicated war, the Boers decided to move deep into the interior and establish their own independent states; one of them was the Transvaal, the other the Orange Free State.

All might have been well if large diamond deposits and, later, rich gold seams, had not been discovered in Boer-occupied lands. Once the flood of speculators and miners began, the situation drifted towards anarchy. Britain was unwilling to be involved in trying to establish law and order in an area over which the Boers wished to remain firmly in control (they denied all political rights to all other settlers, many of whom were engineers without whom the mining could not have been accomplished). However, two factors made it inevitable that before long the British soldier would be called upon to engage in battle. The first was the danger posed by the Zulus, who had an exceptionally martial way of life; the second was that the Boer leader, Paul Kruger, was developing a friendship with Germany from which country the Boers began to obtain large quantities of modern weapons. It became clear that if the Boers were not wiped out by the Zulus, they would use their growing power and wealth to try to eject the British from South Africa. In order to forestall both of these events, the British government decided to annex the Transvaal in 1877. In doing so, they committed themselves to protecting the Boers and settling frontier problems with the Zulus, while establishing a fair and equable system of government within the Transvaal itself.

The Zulu army numbered approximately 50,000 and consisted entirely of fighting men. It was accompanied by women who carried food and all the other supplies; they were nearly as strong as the men and could walk 40 miles (64km) at a stretch while carrying supplies. The Zulu warriors had been trained first by Chaka and then by Cetewayo (their king in 1879) to close in and use the stabbing assegai. Men were not allowed to marry until they had reached a certain age and distinguished themselves in action. When a nation's warriors are not allowed to marry until they have blooded their assegais they tend to be difficult neighbours. In addition to their assegais the Zulus had large numbers of rifles, but preferred to use the assegai as they were poor shots with rifles. The British sent Cetewayo an ultimatum that he should cease border raiding, allow British missionaries to work in Zululand, modify his military system and not mobilise his troops without the consent of the British government. Cetewayo did not reply and accordingly three British columns were despatched to invade Zululand to obtain his agreement to the proposals.

One column, under Lord Chelmsford, advanced from Rorke's Drift to Isandhlwana Hill; the force comprised 1,600 Europeans and 2,500 natives. The core of the force was the 24th Foot, of which a small garrison was left at Rorke's Drift. The 24th Foot were at that time the 2nd Battalion of the Warwickshire Regiment; in 1881 the reconstituted regiment was retitled the South Wales Borderers.

Chelmsford saw no trace of the Zulu army at Isandhlwana and therefore decided to leave six companies of the 24th (about 600 men), some Royal Artillery and a contingent of native troops. He himself went on, to follow up a reconnaissance group he had sent out earlier. While he was away, Isandhlwana was attacked by some 15,000 Zulus.

The battle which followed was an epic event in British military records. The 24th and their companions fought on until every man was killed, but took a fearful toll of the Zulus as they did so. At least 2,000 Zulus were killed in that battle alone and many more wounded warriors dragged themselves away to die on the way home; another 3,500 bodies were found later.

Brecon. The South Wales Borderers have their museum at Brecon. It has a large collection of relics and an audio-visual display. Now amalgamated with the Welch Regiment as the Royal Regiment of Wales, it also shares a museum at Cardiff Castle.

Colonel Chard, who had won a Victoria Cross in 1879 at Rorke's Drift.

No white man saw the end of the Battle of Isandhlwana, but the Zulus later spoke of the steadiness of the Redcoats, how the officers called out and encouraged the men, how the accurate fire from the diminishing Redcoat square eventually caused the Zulus to pause. At that point the 24th taunted them to come on. 'Ah, those red soldiers at Isandhlwana,' the Zulus said, 'how few they were and how they fought. They fell like stones, each man in his place. When they found we were upon them, they turned back to back. They all fought until they died. They were hard to kill; not one tried to escape.' The Zulus knew courage when they saw it. They had plenty of it themselves and they knew that it was the product of regimental pride, comradeship and determination. The 24th had them all.

But the fighting in this war was by no means over. At Rorke's Drift (a ford) there was a base, stores and a hospital. Guarding them was a mixed batch of troops who included one company from the ill-fated 24th (84 men in all), 36 men in the hospital and a company of

🖝

Royal Engineers Museum, Chatham, Kent. Here you can see many interesting regimental exhibits.

Natal Kaffirs commanded by Captain George Stephenson. There were two Royal Engineers in charge of the pontoons on the river: one was Lieutenant Chard, who would win a VC, the other was a private whose fate is unknown. 🖝 Chard heard the news of the disaster at Isandhlwana from two horsemen who had made a lucky escape from the battlefield. Having seen the inevitable outcome they had ridden back to Rorke's Drift to give the alarm. There was little that Chard could do to fortify the camp in the time available, but what little could be done was begun with desperate urgency. Before it could be completed they were attacked by 4,000 Zulus. Those wounded who were able to rise from their beds promptly manned positions on the barricades.

In the open, Chard's scanty force could not have lasted for an hour against such overwhelming odds, but behind the barricades, inadequate though they were, it was a different story. The British soldiers knew that a shot which does not find its target is a shot wasted; they took careful aim and picked off the leading Zulus at a distance. There were too many Zulus for all to be stopped and some forced a way up to the barricades. There, as they tried to climb over the emergency parapets, they were bayoneted or clubbed by rifles wielded by men they could not reach. Sometimes in the hand-to-hand fighting the defenders were able to wrest the assegais from the attacking Zulus and use them against their former owners. The fighting went on all day and into the night. The Zulus set fire to the hospital, which had been emptied and abandoned. Its light illuminated the Zulu army and enabled the garrison, which was blackened, wounded, weary and covered with the blood of themselves or others, to pick targets; the light also prevented the Zulus from rushing the position under the cover of darkness. At midnight the Zulu commander decided that they had done enough and they retired to regroup. They were probably aware that other British forces would be coming up from the rear as reinforcements and if they were counter-attacked in their present state they would all be wiped out. As it was, at least 600 had been killed and many more were wounded.

Lieutenant Bromhead, the commander of the company of the 24th, was also awarded a VC, and three other VCs and many lesser decorations were conferred on the survivors who had fought so heroically. By virtue of fighting behind barricades against warriors whose tactics were based on warfare in the open field, British casualties had been relatively low. Only seventeen were killed and ten seriously wounded.

The British government, appalled by the disaster at Isandhlwana, took decisive action and despatched 10,000 troops. These fought a series of brisk actions before they brought Cetewayo's main force to battle at Ulundi, the Zulu king's well-defended kraal. The British forces included the South Staffordshire Regiment, the Cameronians, the Somerset Light Infantry, the Connaught Rangers, the Northamptonshire Regiment and the

Royal Scots Fusiliers, with the Royal Engineers and Royal Artillery in support. The battle did not last for long, for the Zulus had already been defeated in earlier battles and skirmishes and defeat was turned into a rout by the 17th Lancers and the 1st Dragoon Guards.

One of the most surprising and bizarre events of the war was the death of Louis Napoleon, the French Prince Imperial, who was serving with the British army. When his father, the Emperor Napoleon III, had been defeated by the Prussians in the Franco-Prussian War, he had gone into exile in Britain and his son had joined the British army. After training at the Royal Military Academy, Woolwich (now amalgamated with the Royal Military College, Sandhurst, to become the Royal Military Academy, Sandhurst), he was commissioned into the Royal Artillery. He was successful and popular, both as an army cadet and as a young officer, and when he asked to be given a chance to see action in Zululand it was difficult to refuse him, even though he was the exiled heir to the French throne. It was a disastrous decision, for one day when he was out on a reconnaissance with six other soldiers he was killed by a much superior force of Zulus. Apparently, the party was dismounted when Zulus crept up, unseen, through the long grass. The order to mount was given, but the Prince Imperial slipped and was surrounded by Zulus before his companions realised that he was not with them. He fought desperately, sustaining many wounds on the front of his body. Later his body was recovered and brought back to England. 🖉

Ulundi marked the end of Zulu power but it had unexpected effects. The Boers, who were now free of the fear of being exterminated by the Zulus, turned their hostility against the British. There were only 250 British troops in the Transvaal in December 1880 when the Boers suddenly attacked without warning; all the British were either killed or captured. The Boers assumed that the British would now agree to their independence in the Transvaal, which had been half-promised by Gladstone.

However, the British troops in the Cape and Natal were now mobilised and sent to the Transvaal. The small force consisted mainly of soldiers from the 58th (Northamptonshire Regiment) and the 60th Rifles, but at Laing's Nek it encountered strong Boer opposition and was completely outfought. For the first, but not the last time the British army encountered a remarkably high level of fieldcraft and marksmanship among the Boers. They took advantage of every inch of cover, wore inconspicuous clothing, and used tactics that were appropriate for the ground over which the battle was fought. The word 'Commando', which was used to describe a self-contained Boer raiding column, was now heard frequently. During World War II it was used to refer to British troops raiding German-occupied Europe.

In another battle against the Boers, at Ingogo, British forces were more successful but sustained heavy casualties. Finally, the British commander, Sir George Colley, confronted the Boers at Majuba Hill. His force of 650 men was hopelessly inadequate for the task.

Sandhurst, Camberley. There is a memorial statue to Louis Napoleon on the west end of the New College parade ground at Sandhurst. It was paid for by 25,000 subscriptions from his fellow soldiers and is an impressive and fitting tribute.

Colley, who was killed in the battle, seems to have handled the British troops very inexpertly, but some of his problems may have arisen from the fact that the units under his command were in small detachments which lacked cohesion. Nevertheless, the 92nd (Gordon Highlanders), the Northamptons, the 15th Hussars and members of the Naval Brigade put up a stout fight before they were overrun. The naval contingent lost thirty-six men, nearly half its total strength.

However, although large reinforcements were now on their way to turn the tide of events, Gladstone's Liberal government decided that there was no point in continuing the war. The Boers were granted their demands, which were for an independent Transvaal, and the reinforcements which had arrived at Capetown were promptly sent home again. It would have been wiser to have continued the war, even if the Boers were eventually granted an independent Transvaal, because agreeing to their demands after three humiliating defeats merely encouraged them to believe that they could achieve any ambition they had in southern Africa merely by fighting the British. In short, the seeds of the Second Boer War had been sown and that was a conflict which would involve the British in more defeats and massive expenditure before it was finally won.

The character of the new independent Transvaal was soon revealed. The voting rights of the many British who lived in the Transvaal were curtailed immediately. Raids on adjoining British territories were begun without provocation. Various anti-British laws were passed. All this would have been of no great significance but for the fact that gold was discovered around Johannesburg in 1884 and the Transvaal began to prosper.

Mining and extracting gold required machinery and mining skills; it was not a task which could be accomplished by unqualified people. Engineers came from abroad, mainly from Britain. Soon these immigrants were paying 90 per cent of the total taxes collected, but were refused political rights and schools and courts where English, as well as Afrikaans, would be spoken. But Paul Kruger was adamant: he refused all concessions and used his high revenues to buy arms from Germany. Every attempt to change his mind was rejected. The British reluctantly began to send reinforcements to Capetown and continued to do so even though Kruger warned them that this would precipitate a war. In September 1899 Kruger called up the Boer commandos and informed Britain that unless British troops left the country forthwith he would declare war.

The British may have thought that Kruger was bluffing, but he was not. He followed his ultimatum by invading Natal and the Cape Colony and as he had 90,000 men to confront the 27,000 British who were scattered in small detachments over the vast countryside, the situation looked ominous – in fact it proved to be disastrous. Within two months Kimberley, Mafeking and Ladysmith were under siege and in December 1899, 'Black Week' brought the British forces a

(Opposite page) The Battle of Majuba Hill 1881.

series of disasters which made friend and foe alike wonder whether the British Empire was now finished.

There were many reasons for the early Boer successes. The Boers were superb marksmen, who could shoot accurately even from the saddle of a galloping horse; they used smokeless powder, which made it impossible to tell where their snipers were hidden; they had excellent field artillery supplied by Krupps, and they knew and understood the country in which the fighting would take place.

The British troops, initially commanded by Lord Methuen, fought with great courage but found the terrain and climate almost as much of a problem as the enemy. Among them were the Guards (the Grenadiers, the Coldstream and the Scots), the Northumberland Fusiliers, the Northamptonshire Regiment, the Yorkshire Light Infantry, the Loyals, the 9th Lancers, the Royal Artillery and the Royal Engineers. Lack of cavalry and lack of maps imposed a handicap. The British won a great victory at Modder river but Methuen was wounded and the battle was not followed up immediately when it might have had far-reaching results. Instead, the British next confronted the Boers where they were established in well-sited and expertly concealed trenches at Magersfontein. Here, in addition to the troops already mentioned, were the Black Watch, the Seaforths, the 12th Lancers, the Argylls and the Highland Light Infantry.

Methuen's attack failed; the heaviest casualties were among the Scots, who lost 750 men in comparison with the Boers who lost 250. With the news of the disastrous failure at Magersfontein came the information that there had been another defeat the previous day at Stormberg. But worse was to come.

The force which now set out to relieve Ladysmith and retrieve the situation was much larger, numbering some 21,000 men, of which 16,000 were infantry. It was commanded by Sir Redvers Buller, a veteran of the Zulu and other wars; if Buller could not turn the tide of the war, it was felt, then nobody could. Although Buller looked the prototype of a bloodthirsty old war veteran, he was, in fact, a very compassionate man who paid much attention to his soldiers' welfare and took great pains to avoid casualties. His personal courage was exceptional, but his tactical skills failed to match those of the Boers. In the battle of Colenso on 15 December 1899, the Rifle Brigade, the Durham Light Infantry, the Connaught Rangers, the Queen's and the Devons, all distinguished themselves, but the attempt to break through and relieve Ladysmith failed. The honours of the battle were mainly won by the Irish Brigade, who had 523 casualties, of which 216 were in the Dublin Fusiliers.

This sequence of disasters stirred the British government into action. There were two successful generals available for command. One was Kitchener, who had just returned from the victorious campaign in the Sudan, and the other was 'Bobs' Roberts, veteran of many victories. The choice fell on Roberts, who was despatched forthwith to South Africa to win the war.

Meanwhile, Redvers Buller was still in command of the British forces and the next large battle was at Spion Kop. It lasted from 19 January to 24 January 1900, but it was badly mishandled by the local commander, General Sir Charles Warren, and ended in another Boer victory. However, certain regiments fought with notable courage and distinction, among them the West Yorkshire (now merged with the East Yorkshire to form the Prince of Wales' Own Regiment of Yorkshire), the Middlesex, the Cameronians and the King's Royal Rifle Corps (60th Rifles).

The arrival of Roberts in South Africa transformed the situation. The sieges were lifted and the Boers were systematically cleared from the territories they were controlling. It was obvious that they could no longer hope to defeat the reinforced British in pitched battles, so they changed their tactics. Instead of accepting defeat after the loss of their chief towns, they turned their 40,000-strong army to guerrilla tactics, operating over an area which was as large as England, France and Germany put together. After his victories, Roberts was recalled to become commander-in-chief of the British army in 1900 and was replaced by Kitchener.

Kitchener was an expert and methodical campaigner, but this task was exceptionally difficult. The Commandos were able to rove the country obtaining supplies and assistance from the Boer farms.

British troops on ground reconnaissance in the Boer War.

117

Reconnaissance from the air by Royal
Engineers in the Boer War 1900.

Kitchener decided that as long as they had this supply base, they
would be unbeatable, so he proceeded to remove it. His technique
was the blockhouse system, by which he established small forts
along the railways lines and blocked the intervals between them
with barbed wire. Boer families within the enclosures were therefore unable to supply the roving guerrillas. Many Boer families
were removed from their farms and put in camps at the base; as
they were not used to living in open country and did not understand
sanitation as it is practised in military camps, disease soon broke
out. Many Boers died and the British were blamed for the loss of
life. Their answer was that the Boers were administering the camps
and, in any case; if Kruger had not persisted in continuing the guerrilla warfare, peace would have been signed and the Boers could
have returned to their farms. In 1902 a peace was finally agreed and

its generous terms made friends and allies of many of the Boer leaders.

Kitchener had come to South Africa soon after conducting a similarly methodical campaign in the Sudan. This had been a masterpiece of planning and had involved both railway and river transport. The story begins in Egypt.

CAMPAIGNS IN NORTH AFRICA

Egypt was still nominally a part of the Turkish Empire. However, Britain had a special interest in the stability of Egypt, as the British government had bought a half share in the Suez Canal, which was vital to her route to India and the Far East. Egypt was so badly governed and full of corrupt officials that eventually, after a rebellion there, Britain moved in to restore order. Gladstone reluctantly agreed to the military and naval action, which involved a cavalry action at Kassassin and culminated in the dramatic battle of Tel-el-Kebir (1882) in which the Brigade of Guards, the York and Lancasters, the Black Watch, the Gordons, the Highland Light Infantry and the Cameron Highlanders all fought with distinction. However, after this victory and the restoration of the former Egyptian ruler, Gladstone stated that the British troops would now be withdrawn. Before this could happen, there was a dramatic insurrection in the Sudan, which, at that time, was considered an Egyptian province. The rising was led by the Mahdi, a young religious leader who claimed to be the prophet whom Moslems believe will appear immediately before the end of the world.

The British government sent General Charles Gordon to Khartoum with the brief of withdrawing all the outlying Egyptian garrisons to that city and sitting out the inevitable siege by the followers of the Mahdi, who were known as the Dervishes. Although it was clear that the unfortunate Gordon was doomed if he was not relieved quickly, the British belatedly sent a small force which failed to reach Khartoum before the city fell and Gordon was assassinated. Britain then abandoned the Sudan to its fate.

The Mahdi died soon afterwards and was replaced by the Khalifa, a tyrant who had no claim to spiritual leadership. Although Britain did not wish to be involved in what they knew would be a long and expensive war, the horrific stories told by refugees from the Sudan, and the fact that the Khalifa's Dervish army was a threat to Egypt itself, forced the British government reluctantly to embark on the reconquest of the Sudan to put an end to the tyranny there. For this expedition they put Kitchener in charge. Kitchener, later to be a familiar face on World War I recruiting posters ('Your Country Needs You!'), was a Royal Engineer, as the ill-fated Gordon had been.

Before leaving Gordon we should mention that there are many memorials to that remarkable and controversial personality. A brilliant and courageous soldier, he had also devoted himself to helping young and disadvantaged people.

Gordon School, Chobham, Surrey. A statue of General Charles Gordon, which originally stood in Khartoum, can now be seen at the Gordon School; the school was in fact founded as a memorial to Gordon. Gravesend, where he was once stationed, also has many places bearing his name, and there are Gordon relics at the Royal Engineers Museum in Chatham.

Kitchener set about his task of conquering the Khalifa with that methodical precision for which members of the Royal Engineers are renowned. In order to overcome the supply problem, he organised a system of boats on the Nile and the construction of a railway. Eventually, his army confronted the Dervishes on the plain of Omdurman, just outside Khartoum. His force was heavily outnumbered but better armed, although the Dervishes also possessed modern rifles. Many British regiments distinguished themselves. They included the Cameron Highlanders, the Seaforth Highlanders, the Lincolnshires, the Warwickshires, the Grenadier Guards, the Rifle Brigade, the Northumberland Fusiliers, the Lancashire Fusiliers, the Royal Irish Fusiliers, the Royal Artillery and several locally raised units which consisted mainly of Sudanese. One regiment which became renowned was the 21st Lancers, to which Winston Churchill had temporarily attached himself.

The 21st (now merged with the 17th to form the 17th/21st) made a vigorous charge against a Dervish force but did so unaware that between them and the opponents they could see was a deep ditch, known as a 'khor', in which some 2,000 Dervishes were con-

Belvoir Castle, Lincolnshire. Here you will find the museum of the 21st Lancers.

Officer of the 17th Lancers.

cealed. The 21st numbered 320 and thought they were confronting a similar number. In the ensuing battle, 21 Lancers were killed and 46 were wounded; 119 horses were killed. Although, like the Charge of the Light Brigade forty-four years earlier, the charge of the 21st was an error which caused unnecessary casualties, it split the Dervish forces and made the entry to Omdurman easier.

The campaign had been a triumph of endurance by the British soldiers. In temperatures of about 100°F (38°C), they marched pro-digious distances – for example, 134 miles (216km) in six days, of which 98 miles (158km) were covered in four days. What made this even more remarkable was that the government supplied boots that were so badly made and of such poor quality, that the stitches burst and the soles fell off; many of the men therefore finished the march barefoot.

Army boots improved considerably after the Sudan campaign, and the 'ammunition boot' of World Wars I and II was a success, although it was rather heavy and neither entirely waterproof nor leechproof (in Malaya leeches could get in through the lace holes). The ammunition boot was replaced by the DMB (Direct Moulded Boot), but this proved a failure in the testing conditions of the Falklands War of 1982. The search for the perfect boot continues.

OTHER CAMPAIGNS

The British soldier was called upon to fight in several other wars in the nineteenth century and did so with his customary doggedness and cheerfulness whatever the opponent, terrain or climate. He fought three minor wars in New Zealand against the Maoris, who resented the arrival of British settlers. The cause of many of these disputes was the conflict between the settlers, who believed that potentially rich agricultural land was being wasted, and local people who regarded those lands as their own, even if they were used only for occasional grazing or even were never used at all. Usually the wars came without warning; the local people watched the newcom-ers making incursions on what they considered to be their tribal lands and then suddenly began to raid the farms and kill the settlers.

One of the most complicated of the nineteenth-century wars was in China in 1862, when General Gordon (who was later killed at Khartoum) was in charge of a group of Royal Engineers who were strengthening the defences of Shanghai. The city was threatened by a particularly vicious group of rebels whom the Chinese emperor seemed helpless to suppress. In order to put an end to the threat, Gordon raised a force of 3,500 Chinese peasants whom he led into action, earning it the title of 'The Ever Victorious Army'. Gordon was totally indifferent to personal danger and would advance into the most dangerous situation carrying nothing more lethal than a walking-stick.

Another dangerous war was fought in China in 1900. A Chinese sect, known as the Boxers, suddenly increased in power and influ-

A section of the railway built to support the advance of Napier's expeditionary force through Abyssinia in 1867.

ence, and became violently anti-foreigner. Many foreigners were massacred and those who escaped only did so because they held out in besieged areas. A joint allied force, drawn from countries which had nationals in China and which included British, American, Russian, French, German and Austrian troops, was promptly assembled to raise the sieges and to restore order. Britain's contingent numbered 8,000. The ensuing battles were hard fought and in the two principal conflicts the British contingent played a leading part; in the last of these conflicts they were the first to break through the Boxer lines and to release the besieged groups at Peking. The British force had a special interest in defeating the Boxers, as hundreds of British missionaries and their families, including young children, had been ruthlessly beheaded in public.

Very often British soldiers were sent to overturn local tyrants who were putting to death both their own subjects and foreign nationals who were in or near their countries as traders, missionaries or explorers. One of these tyrants was Theodore, King of Abyssinia, and in 1867 an expedition was sent to attack his stronghold at Magdala. The operation involved a 300-mile (483-km) trek through unmapped and rugged country, followed by an assault on the citadel by means of scaling ladders. The Royal Engineers were to the fore and first over the parapet.

A much longer and more difficult campaign was waged between 1898 and 1904 in Somaliland, which was a British protectorate. The conflict was against a force led by a powerful chieftain who was nicknamed the 'Mad Mullah'. Restoring peace involved an expedition of 6,000 men, but the Mullah himself escaped after a decisive battle and survived to give encouragement to rebel forces in the future.

These campaigns took place on the eastern side of the African continent. Meanwhile, totally different types of campaign were being fought in West Africa. The West African coast was an area where, for most of the century, the British navy had been engaged in suppressing the slave trade and in hunting down the slave traders. In order to stamp out this practice at its source, the Royal Navy had established various bases along the coast. However, the interior of the country was controlled by murderous tribal tyrants, who had formerly sold their own people to the slave-traders and who still raided the peaceful coastal areas for the purposes of plunder and the capture of prisoners.

Napier and his officers in Abyssinia in 1868.

National Army Museum, London. Here you can see King Coffee's bowl.

Royal Signals Museum, Blandford, Dorset. Prempeh's throne, an elaborate chair, can be seen at the Royal Signals Museum.

Three extremely difficult wars were fought against the Ashanti. In 1874, General Sir Garnet Wolseley set off from the coast with a force of 3,500 to conquer King Coffee, ruler of the Ashanti. To reach King Coffee's capital at Kumasi involved an arduous journey through dense jungle and disease-ridden territory. In spite of heavy losses from fighting and disease, Wolseley reached Kumasi and insisted that Coffee should sign a treaty in which he promised to abandon his previous evil practices. Wolseley was a hardened soldier, but even he was sickened by what he found. Coffee had established a sacrificial ground where the earth was literally soaked with blood; this barbarous practice had continued even when he knew that a British force was approaching rapidly. Among the gruesome trophies brought back from this campaign was a huge metal bowl which Coffee was accustomed to fill with the blood of his victims.

Unfortunately, Coffee's successor, King Prempeh, did not feel bound to adhere to the terms of Coffee's treaty and established a similar regime of barbarous practices. Much to the relief of the coastal tribes, who were likely to be carried off by the warlike Ashanti, another British expedition was sent to Kumasi in 1896, this time with only 2,000 men. The difficulties encountered by the British troops were of terrain, climate and disease, for Prempeh was permanently drunk and therefore unable to organise a force to intercept it. The arrival of the British was welcomed by Prempeh's subjects, who lived in fear of his barbaric practices. Prempeh was deported and, in order to preserve stability, a British Resident was installed in his place. British Residents in countries over which Britain had established a protectorate had the onerous task of maintaining law and order, administering justice, raising the local standard of living by trade, and avoiding having their own throats cut by tribesmen who hankered after the old regime.

In 1900 the Resident in Kumasi found himself threatened in this way by the local tribesmen. The town was suddenly besieged by Ashanti warriors and was only relieved when a British force of 1,500 men was sent. After this last battle, Ashantiland was formally annexed to the British Crown and administered by the Colonial Office.

During the campaigns in Africa, several battles were often fought at the same time – for example, while the British expedition was moving against Prempeh in Ashantiland, a brisk war was being fought in Matabeleland, where the rebels were eventually defeated by a force led by Herbert Plumer. He would later take a distinguished part in World War I and become a field-marshal.

Regiments which bear the word 'Ashanti' on their battle honours include the Royal Welch Fusiliers, the Black Watch, and the Rifle Brigade.

A comprehensive list of battle honours won by British regiments between 1662 and 1953 is given in David Ascoli's book *A Companion to the British Army* published by Harrap in 1983.

CHAPTER **9**

WORLD WAR I

WEAPONS AND EQUIPMENT

In the latter part of the nineteenth century and the early years of the twentieth, the lot of the soldier improved considerably. He was accommodated in proper barracks ⟨image⟩ and his food had improved out of all recognition. Discipline was firm but fair, and the cat-o'-nine-tails had long since been abandoned. There was provision for basic education and for sport. In 1898 medical care had been put on an entirely different footing by the formation of the Army Medical Corps. The Cardwell Reforms of 1872 had reorganised the structure of the army and the Haldane Reforms of 1906 continued the work which Cardwell had begun. One of Cardwell's major reforms had been to abolish the purchase of commissions, thus ensuring that all officers had to undergo a period of training at Sandhurst or Woolwich and subsequently to be promoted on merit alone. Haldane reorganised the Yeomanry and Volunteers to form a 'Territorial Army' of fourteen infantry divisions and fourteen cavalry brigades. To provide a supply of partly trained officers, the Officers' Training Corps (OTC) was established in schools and universities.

During this period, there had been a vast improvement in weapons, both large and small. During the Crimean War the Minié replaced the muzzle-loading 'Brown Bess'. In 1868, the Snider, a breech-loading rifle, came into use, and three years later, the Martini-Henry. These powerful weapons fired only single shots and required reloading after each bullet was fired; this was a considerable disadvantage when, for example, a group of Zulus, Ashanti, Dervishes or Afghans was charging at the soldier, who was likely to be unnerved by wild war-cries.

The breakthrough for the infantryman occurred when the Lee-Metford was introduced. The Lee-Metford of 1888 had the .303 calibre, bolt-action and magazine which later became familiar to millions in the Lee-Enfield, which replaced it in 1895. The bolt action meant that when the bolt was withdrawn it ejected the last (fired) round and allowed another to be pushed up from below by a spring in the magazine. As the bolt went forward and rammed home the round, it was ready for firing again. Trained soldiers could soon fire Lee-Enfields (which could take up to ten rounds in their early replaceable magazines) with such steadiness and precision that their opponents were likely to believe that they were facing automatic weapons. The SMLE (Short Magazine Lee-Enfield) was the standard infantry weapon through both World War I and World War II, although in the former it was supplemented by pistols, mortars and hand-grenades, and in the latter by Thompson ('Tommy') sub-machine guns, Sterlings, Stens, Patchetts, Brens and anti-tank

TERRITORIAL ARMY.

SMART MEN WANTED

Field Telegraphy

FOR THE FIELD COMPANIES AND TELEGRAPH COMPANY OF THE

Double Tool Cart

ROYAL ENGINEERS

(2ND LONDON DIVISION)

Headquarters, 67, COLLEGE STREET, CHELSEA, S.W.

Pontoon Bridging

Pontoon Waggon

Surveying

Spar Bridging

Terms and Conditions of Service in the 2nd LONDON DIVISIONAL ENGINEERS.

Colonel E. T. CLIFFORD, V.D., Commanding.

❋ Headquarters: 67, COLLEGE STREET, CHELSEA, S.W. ❋

The Instruction given in this Arm of the Service is of great value to men who are engaged in the engineering, building and kindred trades.

TRADES REQUIRED For the Field Companies.	MOUNTED BRANCH.	Age and Standard of Measurement.	TERMS OF SERVICE.	DRILLS TO BE PERFORMED.	ANNUAL TRAINING.	PAY & ALLOWANCES During Annual Training.
Carpenters & Joiners, Masons, Bricklayers, Blacksmiths, Gas-Fitters, Plumbers, Fitters & Turners, Painters, Engine Drivers, Farriers, Wheelwrights, Shoemakers, Tailors, Surveyors, Slaters, Plasterers, Harness-makers, Coopers, Draughtsmen, Electricians, Printers, &c.	Young Men, used to Horses. ARE REQUIRED for Service IN THE MOUNTED PORTION OF THE Field Companies AND Telegraph Company	AGE 17 to 35. HEIGHT 5 ft. 2 in. and upwards. CHEST MEASUREMENT 33 inches	MEN are required, on Attestation, to engage for a term of 4 years, and on completion of this period may re-engage at intervals until reaching the age of 44. Non-commissioned Officers of the rank of Sergeant and upwards, may continue to serve until 50 years of age.	FIRST YEAR Engineering Drills - 35 Infantry or Riding Drills - 30 OTHER YEARS Engineering Drills - 10 Infantry or Riding Drills - 5 PRIZES are given annually for General Efficiency, Engineering, Shooting, Riding and Attendance.	Annual Training must be performed at Camp for not less than 8 or more than 15 days unless leave of absence has been obtained from the Commanding Officer. Applications for leave must be supported by Medical or Employers Certificate.	Pay and Allowances will be at Army Rates, as under :— Corps.-Sergt.-Major, per day		For Telegraph Coy Telegraphists & Wiremen, & Instrument Repairers.					

Application for attestation should be made to the Sergeant-Major, or either of the Sergeant-Instructors at Headquarters.

❋ GOD SAVE THE KING. ❋

The Sir John Fowler exhibit at Blandford.

weapons. The SMLE remained in service until 1968 and was then replaced by automatic weapons, the first being the SLR (Self-Loading Rifle), of Belgian design. The SLR has now given way to the SA-80.

Soldiers are always told 'Your best friend is your rifle' and in time of danger it certainly is. However, when a soldier is required to clean it to perfection for parade-ground inspection and is punished for having a speck of dust or a finger-mark on an otherwise gleaming surface, the friendship becomes a little less close. Although longer than modern weapons, and heavy, the SMLE was considered by many to be an ideal weapon. It was accurate, reliable and could kill at distances of up to a mile. After becoming obsolete it had another lease of life as a marksman's rifle, usually after it had been fitted with an internal sleeving which narrowed the calibre of the barrel.

(Opposite page) A recruiting poster for Territorial Army volunteers in 1907.

Since 1860, the infantryman had also had the benefit of the Gatling, a multi-barrelled machine-gun, which was turned by a crank and which fired at each turn of the handle. The Gatling was replaced by the Maxim after 1891. The Maxim, a genuinely automatic weapon, could fire 600 rounds a minute. During World War I the Maxim was replaced by the Vickers Machine-Gun, which remained in service until 1968, although in World War II it was only one of several available machine-guns, the Browning being another favourite. The standard weapon today is the GPMG (the General Purpose Machine-Gun).

National Army Museum, London; Weapons Museum, Warminster. A large selection of weapons may be seen at these two museums. However, a visit to the Weapons Museum requires a prior appointment with the curator.

The cleaning and polishing of weapons and equipment to a degree well above the point of efficiency, and sometimes to a point at which efficiency is impaired, seems to have derived from the centuries in which soldiers were stationed in India. Although, when campaigning, troops were out in the heat of the day, it was considered healthier in peacetime that they should do their drilling and other work in the cool of the morning, or even in the evening, and stay in barracks when the sun was at its most fierce.

To keep the soldier busy in barracks, remarkable standards of cleanliness were devised. His topee (sun helmet) was scrubbed and blancoed, while his leather equipment was polished to a mirror-like surface. The humble boot required the most attention of all. When issued, boots were made of greasy leather and one pair was allowed to be kept in that state for training. The other pair was scrubbed and polished until the toecap literally reflected the inspecting officer or NCO; even the studs in the sole were polished, as were the brass tags on the bootlaces. The process of burning and burnishing boots usually ruined them because it dried them and made them crack, but the chore of shining boots, to this day, is considered to be the basic parade-ground requirement of a smart regiment. The tin in which his boot polish was supplied would have the paint laboriously scraped off and then be polished to such a degree that the soldier could shave by it. But at least he did not have to powder his hair as his predecessors had done. When webbing was substituted for leather, blancoing (which could mean green, black or white) was an equally painstaking performance. Puttees could not be polished or blancoed but could land him in trouble if they were not put on with each fold at a meticulously accurate distance from the next. In theory, puttees were a form of bandage designed to protect the ankle and lower limbs from scratches, snake-bites, etc, but in practice they became yet another article of clothing that the soldier had to keep in perfect condition or be punished with extra drills or extra kit inspections.

Soldiers no longer need to have every moment of their time filled with chores, but high standards of turnout from recruits are considered necessary to the training of a smart soldier. And it is, of course, true that a man, or woman, who has been taught to attain perfection in the appearance of such familiar articles as everyday equipment should be able to apply the same painstaking attention to other basic military skills. Added to this is the fact that most soldiers take enormous pride in being able to reach very high standards in the care of personal kit and weapons. No soldier describing another unit can find more contemptuous words than 'A scruffy lot', however unjust the accusation may be. 'Good soldiers die with their boots clean': the statement may not often be true but it indicates an essential principle.

When Britain became involved in World War I on 4 August 1914, the British soldier was well trained, well disciplined, properly administered, but badly paid. Standards of marksmanship were high: so rapid and accurate was the fire of the British infantry that the Germans who encountered that wall of steel indeed believed that it must be coming from automatic weapons. The same high standards of shooting prevailed in the Royal Artillery, which had 18-pounder guns, 4.5-inch howitzers, and 60-pounders which had a range of 9,500yds (8,700m). The Horse Artillery, which was more mobile, as may be seen when the RHA perform at tournaments and similar occasions, had 13-pounders. All these guns gave magnificent service in World War I.

"Well, if you knows of a better 'ole, go to it."

A Bairnsfather cartoon – humour in adversity.

YPRES AND PASSCHENDAELE

Britain had no direct quarrel with Germany in 1914 but was drawn into the conflict because, in its strategic move to defeat France, the German army invaded neutral Belgium. Belgium had become an independent country in 1839 and that independence had been guaranteed by Britain. Much to the Kaiser's surprise, the British honoured what he called 'a scrap of paper'; he was even more surprised when Britain's small regular army, which he described as 'contemptible', frustrated his plans to sweep through to Paris. (The Germans actually reached a point 14 miles (22km) from it.)

The German strategy was based on the 'Schlieffen Plan', which was named after a general who died even before the war broke out. It involved the main German armies pivoting upon Metz and coming around in a huge arc (violating Belgian neutrality on the way) until they enveloped Paris and crushed the French armies against other German forces invading further south. However, the plan went astray when they were opposed by French forces at Charleroi and British forces at Mons. Although both the French and the British retreated, they had done enough to disrupt the German plan. In their frustration, the Germans failed to capture the Channel ports, which had been one of their original intentions.

By 5 September, a month after their invasion of Belgium, the Germans were on the Marne and Paris seemed to be in their grasp. At that moment the French General Joffre launched a counter-attack which threw the over-extended Germans into confusion. The Germans retreated and now tried, too late, to seize the Channel ports. Britain was determined that this should not happen and put up a strong resistance around the town of Ypres. Meanwhile, both sides began hastily to dig a line of trenches to hold the terrain they

British 9.2in heavy howitzer for use in trench and siege warfare.

were occupying at that time. The trenches soon extended from the Channel to Switzerland: 400 miles (644km) in all.

At first the trenches were shallow but, as the months passed, they were deepened and equipped with communication trenches from behind and barbed wire in front. In the next three years there were desperate attempts by both sides to drive a wedge into the opposing line and then to make a breakthrough. However, the successes achieved were minimal in relation to the enormous cost in lives of gaining a few hundred yards of ground. Battle honours were being won while the war was only a few months old. At Mons, the Royal

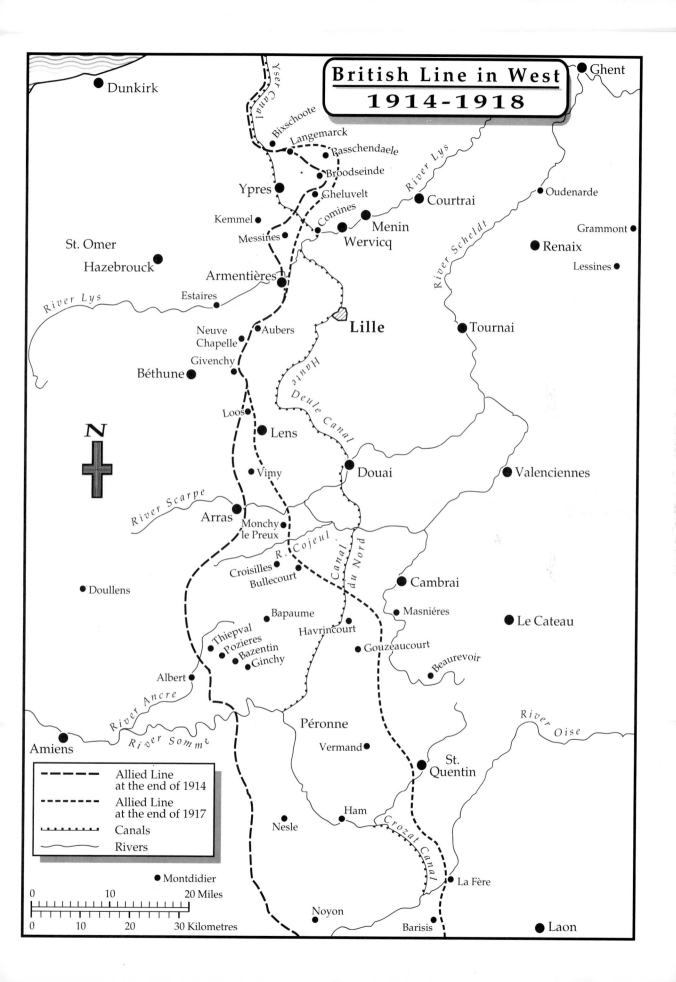

British Line in West
1914-1918

Dunkirk

Ghent

Bixschoote
Langemarck
Passchendaele
Broodseinde
Ypres
Gheluvelt
Comines
Kemmel
Menin
Messines
Wervicq

River Lys
Courtrai
Oudenarde

Grammont
Renaix
Lessines

St. Omer
Hazebrouck

Armentières
Estaires

River Lys

Neuve
Chapelle
Aubers
Givenchy
Béthune

Lille

River Scheldt

Tournai

Haute Deule Canal

Loos
Lens

Douai

Valenciennes

N

Vimy

River Scarpe

Arras
Monchy
le Preux

R. Cojeul

Croisilles
Bullecourt

Canal du Nord

Cambrai
Masniéres

Le Cateau

Doullens

Bapaume

Thiepval
Pozieres
Bazentin
Ginchy

Havrincourt

Gouzeaucourt

Beaurevoir

Albert

River Ancre

Péronne

River Oise

Amiens

River Somme

Vermand

St.
Quentin

Allied Line
at the end of 1914

Allied Line
at the end of 1917

Canals

Rivers

Ham

Crozat Canal

Nesle

Montdidier

0   10   20 Miles

0   10   20   30 Kilometres

La Fère

Noyon

Barisis

Laon

Fusiliers, the Royal Northumberland Fusiliers, the King's (Liverpool), the Suffolks, the Norfolks, the Lincolns, the Royal Scots Fusiliers, the Cheshires and the Queen's were among the many regiments which distinguished themselves. At Le Cateau, the Warwickshires, the Hampshires, the East Lancashires, the East Surreys and the Duke of Wellington's put up a dogged defence. One of the most extraordinary achievements was that of the 2nd Worcesters (now amalgamated with the Sherwood Foresters to form the Worcesters and Foresters). On 29 October 1914 the Germans were forcing through an attack towards the town of Ypres. They had taken the important village of Gheluvelt, which they now occupied with 1,200 men. At that moment, the 2nd Worcesters, who were reduced by casualties to 350 men, was launched at them by their temporary commander, Brigadier-General Fitzclarence, VC. Unbelievably, the fierceness of the attack at the point of the bayonet sent the Germans hastily into retreat and the gap in the British line was sealed. Fitzclarence was killed a few days later.

Regimental Museum, County Museum, Worcester. Here you can see the relics, weapons and uniforms of the 2nd Worcesters.

The area to the west of Ypres, which became known as the Ypres Salient, became one of the most devastated battlefields in the history of warfare. The Germans held a semi-circular ridge 7 miles (11km) from the town of Ypres and from this superior altitude could shell Ypres and the British forces in the Salient at will. On the crest of the hill was the small village of Passchendaele, which became the object of massed British infantry attacks over the next three years. The battlefield had been Belgian pastureland, which lay at sea-level and which was drained by a complicated system of interlocking canals. Once the artillery barrages started, the whole region was transformed into a morass that was pitted with huge shell-holes. By the time the Allies reached Passchendaele (which was captured by Canadians), a quarter of a million British, Australian and Canadian soldiers had become casualties; 90,000 of these had disappeared without trace in the liquid mud and shell-holes. The Germans had attempted a breakthrough in the spring of 1915 by using gas for the first time in the war. The way to Ypres lay open but they failed to realise it and missed their opportunity. From then on the British used gas too, and the whole of this evil battlefield reeked of mustard gas, corpses and general decay.

In July 1917, Douglas Haig, the British Commander-in-Chief, decided to launch a devastating attack which would carry the British forces to Passchendaele Ridge within days. As a preliminary, the Messines Ridge, to the south of Passchendaele, where the Germans were very strongly entrenched, was to be blown up. The operation, which was managed by deep mining in great secrecy, involved a million pounds of high explosive and was heard even in London. However, it was not a decisive blow and when the British attack went in on 31 July 1917 it was held up by resolute Germans, powerful defences and relentless rain. Perhaps the worst enemy on that battlefield was the almost continuous rain. As a result, when Passchendaele was finally captured in November 1917, the winter,

The battlefield of Passchendaele – a graveyard for men, horses, and tanks.

with the prospect of even worse weather, was at hand, and the desperately hard-earned victory could not be exploited.

The heroism and stoicism of the soldiers on this battlefield defies belief. Numerous VCs were won, but there must have been many more which were earned without being observed: a recommendation for a VC requires three independent witnesses in conditions in which there is a 90 to 100 per cent chance of death. It was easier to find the latter than the former. In the entire war only one man won the Victoria Cross twice. He was Captain N. G. Chavasse of the Royal Army Medical Corps, who was attached to the King's Regiment. In earning the second on 31 July 1917 he was wounded so badly that he died soon afterwards; as well as his two VCs he had also been awarded an MC. Only three men have ever won the VC twice and two came from the RAMC.

It would be impossible to list all the honours that were won on the Passchendaele battlefield – space does not permit it. But there are few museums which the reader can visit which will not have some record of the incredible deeds of courage and endurance that were performed on a battlefield whose horrors have never been exceeded.

Royal Army Medical Corps Museum, Aldershot. Here you will find many displays relating to the RAMC.

LOOS

The Allies tried various means of breaking through the lines at what appeared to be the most vulnerable sections in 1915. In March of that year a British attack at Neuve Chapelle was initially a great

Situation Shortly Vacant.
In an old-fashioned house in France an opening will shortly occur for a young man, with good prospects of getting a rise.

Humour, but revealing the inhumanity of war: the domestic clock and candlestick from a demolished home.

success. British infantry overran the German trenches which had been cut off from their support areas by devastating Royal Artillery barrages. However, before the British attack went into the second phase, there was a delay of five hours while the British commanders assessed the meagre reports and debated what to do next. That five hours was vital to the Germans who used it to reorganise their defences. When the second phase was pressed home the losses were enormous. Not least of the causes of the disaster at Neuve Chapelle was the shortage of shells; nobody had visualised the enormous number which would be required and in some sectors artillery batteries were restricted to a few rounds a day. When he was subjected to heavy German barrages, to which there was little reply, the British soldier, who seemed to be able to preserve his whimsical sense of humour under the most impossible conditions, would call out, 'Give over, Jerry, a bit. Our gun's broke.'

Shortage of shells also contributed to heavy British losses in subsequent attacks at Aubers Ridge and Festubert. However, the Germans, who at this stage had no shortages to contend with, were no more successful when they tried to advance.

In September 1915 the British army launched another massive offensive, this time at Loos, between La Bassée and Lens. General Haig, who was subsequently blamed for the heavy losses in this offensive, had originally opposed it but had been overruled by his French superiors, Joffre and Foch. Haig contended that the supply of shells was still inadequate (although desperate attempts were being made to increase production in Britain) and that the terrain was totally unsuitable for massed infantry attacks. Nearly 60,000 casualties were incurred at Loos, some 2,500 of them in the first twenty-four hours. Even so, the battle came within a fraction of being a victorious breakthrough. If it had, and been followed up, the war would have ended that year and there would have been no Russian collapse and consequent revolution, no need for American intervention and no devastating financial burden on all the combatants.

Loos was the first battle in which the British used gas, and the experience proved how difficult and unpredictable this new weapon was. First, the large and heavy cylinders had to be carried up to the front line in the dark in order to preserve complete secrecy. Second, when the gas was released it depended on a favourable wind to carry it over the enemy trenches. In the event, some of it blew back on to British lines. In spite of this, British troops captured 8,000yd (7,300m) of German trenches and in places penetrated up to 2 miles (3km). This was all the more remarkable because two divisions, the 15th (Scottish) and the 47th (London), came from 'Kitchener's Army'. Kitchener had been the only person in 1914 to realise that the war would last for years and would require at least a million men. He therefore set about encouraging volunteers. What they lacked in training these volunteers made up for in enthusiasm. Unfortunately, many of the men killed in the early battles were potential officers; later in the war, when so many officers had been killed, it became necessary to replace them with men who previously would not have been considered suitable. Most of the 'Old Contemptibles', as the pre-1914 regular army took a pride in being termed, had been killed in resisting the 1914 German attacks and even raw recruits like the 47th had sometimes had a taste of battle with little more experience than firing a few practice rounds. They included such unwarlike names as the Post Office Rifles, the Civil Service Rifles, and the Poplar and Stepney Rifles; nevertheless, they fought in the best traditions of the British army. The infantry regiments at Loos included in their ranks many men who should not have been fighting in that capacity at all. They were medical students and chemists, engineers and scientists, and many others with technical experience which would later be vitally important in other arms such as gunnery, medicine, signals and ordnance; many were killed.

One of the regiments to suffer when the gas blew backwards was the 7th King's Own Scottish Borderers. Inadequate gas masks made breathing almost impossible, but if they were removed, the soldiers were asphyxiated. Their predicament caused such confusion that the line did not move forward until the sound of the pipes was heard. These were played by Piper Laidlaw, who marched up and down the parapet, and was subsequently awarded a VC. Another gallant piper was Pipe-Major Robert Mackenzie, then aged 60. Although he was wounded, he kept on piping; then he was killed. Mackenzie was not the oldest man in the battle. That seems to have been Major William Porter of the Royal Sussex Regiment. He was 66 in September 1915 and was wounded on the second day; he survived until he was 80 years old.

> **King's Own Scottish Borderers Museum, Berwick-upon-Tweed.** Holds the regimental relics, weapons and uniforms.

The casualty figures at Loos were appalling, but they did not daunt the survivors. The 10th Gloucesters lost all but 60 men; the 8th Devons were still attacking when their numbers were down to 2 wounded officers and 100 men. The Durham Light Infantry lost 17 officers, which included the CO, the adjutant and all the company commanders. The commanding officer of the Buffs was 61; he was killed with 12 other officers.

An unknown officer described an attack as follows: 'I could not have imagined that troops with a bare twelve months' training behind them could have accomplished it. The sight of these seemingly unconcerned Highlanders advancing on them must have had a considerable effect on the Germans. I saw one man whose kilt had got caught in our wire as he passed through a gap: he did not attempt to tear it off but carefully disentangled it, doubled up to his correct position in the line and went on.'

The price of the troops' fortitude was high. In the 45th Brigade, the following deaths were recorded: the Royal Scots – 13 officers and 609 soldiers; the Royal Scots Fusiliers – 10 officers and 598 soldiers; the Cameron Highlanders – 13 officers and 600 soldiers, and the Argylls – 15 officers and 611 soldiers.

The high casualty rate extended over the entire battlefield, which included units from all parts of the British Isles. Loos was one of the most important battles of the war and certainly one of the bravest and bloodiest but, because it came early, it was subsequently overshadowed by the great battles of the Somme, Verdun, Passchendaele, Arras and Vimy Ridge.

THE 'SIDESHOWS'

In 1915 public attention was also drawn to other theatres. On 25 April an expedition landed at Gallipoli. The aim was to advance from there to Constantinople (Istanbul) and to knock the Turks out of the war. The Turks were in alliance with Germany and controlled valuable strategic territories such as Palestine (Israel), Syria and Mesopotamia (Iraq). The planned attack at Gallipoli was Winston Churchill's brainchild. He believed that a successful attack in that area could be the first step in turning the German flank, linking up with Russia and thus breaking the deadlock of trench warfare on the Western Front. In the event, the Turkish/German defences were too strong for the expedition to force a way inland and the battle settled down into another form of trench warfare in which the Allies were at a great disadvantage. Eventually, after sustaining 100,000 casualties, the expeditionary force was withdrawn. The Anzacs (Australian and New Zealand Expeditionary Force) fought with great courage but suffered enormous casualties and hardships. By the beginning of 1916, the evacuation, and the failure, was complete. It was, however, a costly operation which came very close to success.

In the appalling conditions many regiments distinguished themselves. Perhaps the most famous was the Lancashire Fusiliers (now part of the Royal Regiment of Fusiliers) who won six VCs before breakfast.

Another distant theatre was Salonika (Thessaloniki) in northern Greece. In October 1915 Germany's ally Austria had invaded Serbia. British and French divisions were promptly sent to Salonika with a view to assisting Serbia, but the Serbs were overwhelmed before military help could reach them. The Allies would have been wise to have evacuated their force and used it elsewhere but, instead, held it in place to keep a watchful eye on Greece, which looked like falling under German influence. Salonika was an extremely unhealthy base and the life of the soldiers stationed there was more likely to be cut short by disease than military action.

The situation in Mesopotamia was very different. At this time Britain obtained her vital oil supplies from Persia and, in order to protect these supplies, it was considered advisable to control Mesopotamia. However, Mesopotamia was part of the Turkish Empire and the Turks maintained a substantial force there. Although Basra, on the Persian Gulf, had been captured as early as November 1914, plans to advance further into the country soon ran into considerable difficulties. The British General Townshend

Lancashire Fusiliers Museum, Wellington Barracks, Bury, Lancaster.
This museum holds regimental mementoes and relics of Wolfe.

aimed for Baghdad but encountered superior Turkish forces and fell back to Kut. Here he was besieged and, after four months, had to capitulate. The Turks considered that Townshend's army had fought magnificently before it was forced back to Kut, but this did not influence them to treat their prisoners with anything but extreme harshness: of the 9,000 men taken, 5,000 disappeared without trace. However, the Turks were not allowed to retain Kut for long. A new and stronger force was sent to recapture it, which was accomplished on 25 February 1917. The British force, which contained many units from India, pressed on, and soon won Baghdad.

The Middle East campaign again illustrated the versatility of the British soldier and also did much to develop the expertise of the medical services. In the wet season the weather was icy cold and men battled with flooded rivers; in the summer the temperature was 102°F (39°C), and troops lived in tents which became as hot as furnaces. For long periods there were no comforts, not even cigarettes, and rations were short. Dust blew around in clouds and the soldiers were constantly bitten by every variety of insect. Regiments which carry the words Tigris, Kut or Baghdad on their battle honours have good cause to be proud of their record under those conditions. They include the Royal Leicestershire Regiment (now part of the Royal Anglian), the Royal Welch Fusiliers, the South Wales Borderers (now part of the Royal Regiment of Wales), the East Lancashires, the Dorsets (now joined with the Devons), the Black Watch, the Oxford and Bucks, the Manchesters and the Wiltshire Regiment.

Duke of Edinburgh's Royal Regiment, Salisbury. Relics, displays, weapons and uniforms of the regiment.

It became the custom to refer to these hard-fought campaigns as 'sideshows', but many of them were more gruelling than battles on the Western Front because there was no home leave, no relief from the climate and men were not taken out of the line and rested as they were in France. In France, of course, a man might not survive long enough for relief or leave, but if he did his situation was better than for the soldiers who were committed to Gallipoli, Mesopotamia or certain other theatres.

Another 'sideshow' was Palestine, which at that time was a Turkish province. British attempts to conquer that country met with no success until General Allenby took over command in July 1917. Allenby was a brilliant strategist, who was much respected by his men (who nicknamed him 'The Bull') and who had already been a successful general in France. Palestine was a cavalry campaign in which Allenby outmanoeuvred the enemy, which included both German and Turkish forces. He had a number of excellent Anzac units and a number of British Yeomanry regiments. In Palestine troops encountered the problems of heat, dust and disease. The campaign was brought to a successful conclusion.

THE BATTLE OF THE SOMME

In order to present a broad picture of the progress of the war on other fronts, we left the Western Front at the end of 1915. In 1915, both the French and the Germans tried heavy attacks on each other's line but both took very high casualties. Neither side appeared to have learnt anything from the enormous losses they had incurred in conventional infantry attacks. The normal pattern was to give warning of the impending assault by a long and continuous artillery barrage. During its closing stages, soldiers would be massed in the front-line trenches, waiting to 'go over the top' when it stopped. In theory, the shelling would have reduced the enemy strongpoints, front-line trenches and machine-gun posts to rubble; it would also have destroyed the barbed wire. In the event, the barrages were far less effective than was expected. Often they failed to cut the wire, which then had to be tackled by soldiers with wire-cutters operating under almost impossible conditions. Although the barrage might have reduced the front lines to rubble, soldiers would emerge from deep dug-outs and set up their machine-guns; these would take a fearful toll on the advancing infantry which would be totally exposed in the open space of 'No Man's Land' and often held up by uncut barbed wire.

Infantry platoons were led by officers and, as these were likely to be killed in the first wave of enemy bullets, the men following behind would have little idea of what to do except to try to forge ahead. Once an attack began, it was almost impossible to know what was happening; one regiment might be making progress while another might have been virtually wiped out before it had gone more than a few yards. Success sometimes brought its own penalties. A company or platoon which advanced too far was likely to be surrounded by the enemy; if they were not killed, the men could then be taken prisoner. This situation was the same for both sides, although circumstances favoured the Germans, who had

A log road in Flanders in 1917, built on liquid mud.

more machine-guns, better prepared positions and, at this stage, more trained troops.

No commander likes to see the flower of his country's youth being slaughtered in mass attacks, but when countries are locked together in a desperate struggle, high casualties are inevitable. During World War I, the British soldier proved that he was the best in the world at both defence and attack. 'British' was then used to describe all the soldiers from the British Empire and included Canadians, Australians, New Zealanders and South Africans.

The Germans tried to make a decisive strike against the centre of the French line on 21 February 1916 by launching a devastating offensive against Verdun on a 21-mile (38-km) front. Having occupied the French front-line trenches, they assumed that they would soon be in possession of the city. They could not have made a greater mistake. The French fought back and the grinding battle lasted until 15 December, both sides losing hundreds of thousands of men. The Germans lost 120,000 men in one week.

Relief for the French came when Haig began a massive offensive on the Somme on 1 July 1916. The assumption that the German defences would have been destroyed by the preliminary eight-day bombardment proved tragically mistaken. When it lifted, the Germans emerged from their deep dug-outs (whose existence had been unsuspected) and mowed down the British infantry in thousands: 40,000 were said to have fallen in the first three hours. By the time the Battle of the Somme had dragged to its end, another 450,000 men had died. The Battle of the Somme was designed to help the French defence of Verdun, and certainly it did so, but the cost was almost unimaginable. The brunt of the Somme battle fell almost entirely on the British, although it was meant to be com-

Royal Ulster Rifles Museum, Belfast. The 36th (Ulster) Division achieved its objectives at enormous cost on 1 July 1916. At the Royal Ulster Rifles Museum a full range of regimental exhibits are on display.

'Little Willie', the first tank (1915), now on display in the Tank Museum, Bovington, Dorset.

Mark I Tank (1916), the only surviving example of the first tank used in action. It can be seen at the Tank Museum, Bovington.

bined with a French attack. However, the French had now been so badly mauled that they were incapable of any more large-scale offensives and the British had to take up the main burden of the war in France. In 1917 the French army mutinied and, as a result, parts of the French sector were temporarily left unmanned; fortunately, the Germans were not aware of the opportunity this presented.

The Battle of the Somme had begun at 7.30am at the beginning of a bakingly hot summer day. Each soldier was carrying 66lb (27kg) of equipment and advanced shoulder-to-shoulder with his comrades. The weight made it impossible for him to move at more than a slow walk. The aim of burdening the soldier with equipment weighing nearly half as much as his own weight was to ensure that when he arrived at his objective he would have enough equipment to settle in and hold it. Although many soldiers were killed as they left the trenches, or on the wire, some got through and gained ground up to a mile. One of the greatest tragedies of this battle was the slaughter of so many close friends who had joined up together, trained together, and then all died together. Among these were the 'Bradford Pals', the 16th and 18th battalions of the West Yorkshire Regiment. Sadly, the Bradford Pals were not unique. In other towns whole streets of close friends joined up and died together. ☞

The Bolling Hall Museum, Bradford. Commemorates the 'Bradford Pals'.

In the final stages of the Battle of the Somme a new weapon appeared – the tank. Plans for an armoured fighting vehicle had been rejected by British, French and German armies in the pre-war years, but in 1915 British tank development was encouraged by Winston Churchill and in consequence forty-nine tanks were brought into action on 15 September 1916. They were huge, cumbersome machines, with a speed of 4mph (6kph), but their appearance on the battlefield was as bad for German morale as it was good for the British. ✍

ARRAS

On the Somme the tanks were not used sufficiently intelligently to have much influence on the battle, which was then coming to a close. They were in action again at Passchendaele the following year, but the mud turned the area into a 'tank graveyard'. Their finest hour during World War I was at Cambrai in November 1917, when 381 tanks broke through the German lines and made a gain of 3 miles (5km) for the loss of less than 100 men. At Passchendaele, thousands of lives had been sacrificed for lesser gains. However, most of the gains at Cambrai were quickly lost again in German counter-attacks and not until the following spring did tanks really show what they could do. On 26 March, seven Whippet tanks (with a top speed of 8mph (13kph) and a range of 80 miles (129km)) broke up a German attack, killing nearly 400 men for the loss of 12.

In early 1918 the German army massed itself for one final devastating attack with which they planned to secure victory in France.

Army Tank Museum, Bovington, Dorset. One of the finest museums in Britain is the Army Tank Museum. There the visitor will see early tanks and many of later design from armies from all over the world. At Bovington, the visitor can get close enough to the tank to understand what tank warfare was really like.

Stretcher bearers at Pilckem Ridge, Belgium in 1917. In spite of the horrific difficulties they saved many lives.

The collapse of Russia and her virtual disappearance from the war meant that Germany could bring back troops from that front and use them in the west. Realising that the French were in no condition to hold it if the attack fell in their sector, the British relieved them by taking over an additional 25 miles (40km) in the most likely spot for the German assault. The American contribution to the war was just beginning, but at this stage the American army was inexperienced and was having to be supplied with its weapons and aircraft by the Allies; however, the fact that American soldiers would soon be arriving in vast number spurred the Germans to launch a massive all-out attack to end the war before they arrived.

The blow fell on 21 March: 74 German divisions, supported by 6,500 guns, fell on to 30 British divisions between the Oise and Arras, the heaviest impact falling on the 14 divisions of General Gough's Fifth Army. Although Gough's men fought desperately, they could not hold the German assault and the Germans eventually made gains of up to 60 miles (96km). A feature of the resistance was that every soldier, whatever his trade, had to fight as an infantryman – cooks, transport drivers, gunners, signallers, engineers and mechanics all played a part, one which no one would previously have thought them capable. Among the many regiments to distinguish themselves by their dour fighting qualities were those in the 46th Division: the Lincolns, the Leicesters, the North and South Staffordshire, and the Sherwood Foresters. Australians and Ameri-

(Opposite, above) Early Crossley and Lanchester armoured cars.

(Opposite, below) A 1917 tank.

Aircraft on display at the Imperial War Museum, London.

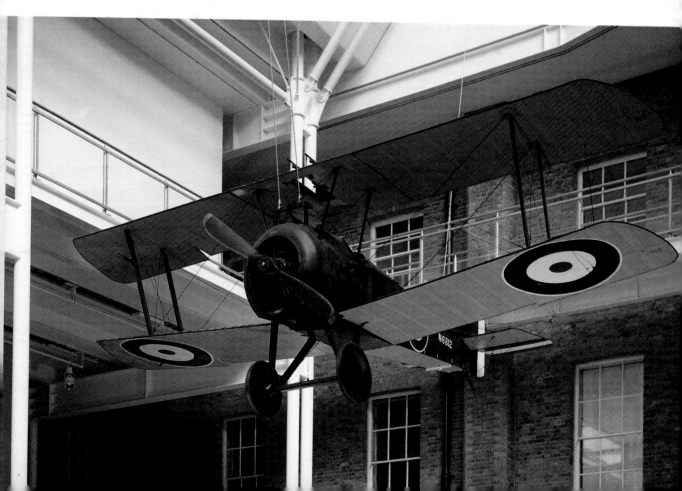

cans also fought back doggedly. For the 2nd Devons, who were no strangers to desperate battles, this was probably 'their finest hour'. On 27 May the Devons were holding the last trench north of the Aisne at a point called Bois des Buttes. They numbered 28 officers and 552 NCOs and men – that is, little over half the strength of battalions in other theatres. Unluckily for them – or perhaps they would have preferred to say 'luckily' – they stood in front of a German division which desperately needed to continue the advance. They were pounded mercilessly with all the artillery the Germans could range on to them. They hoped to hear that reinforcements were on their way to help them, but they only learnt that units on either side of them had given way. They fired their Lewis (machine)

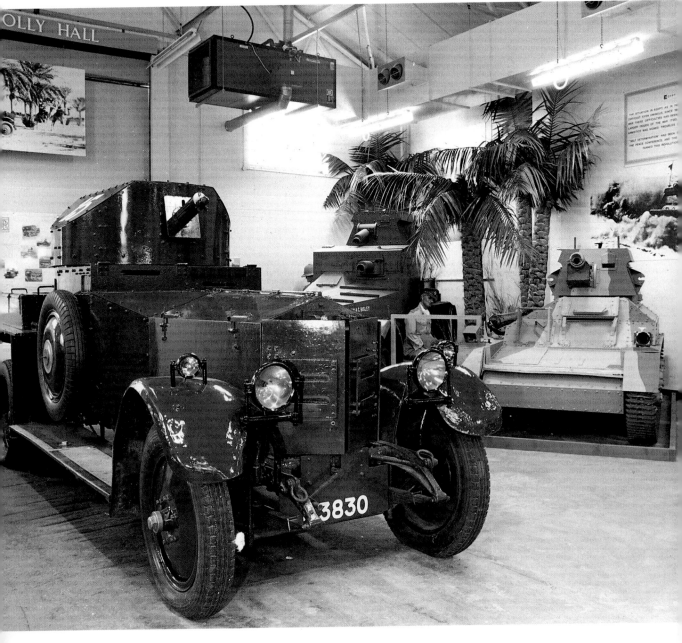

Rolls Royce 1920 armoured car. This can be seen at the Tank Museum, Bovington.

guns until the barrels were too hot to touch. In places the Devons were overrun, but they fought on in small batches, making it impossible for the Germans to break through. As the Devon officers and NCOs fell, their juniors took over their posts. 'B' Company was wiped out altogether, while 'A', 'C' and 'D' companies were holding on desperately. The colonel told the survivors that their last chance of escape had gone, as the Aisne bridge had been lost. 'Your job for England, men, is to hold the blighters up as much as you can and give our troops a chance on the other side of the river. There is no hope of relief. We have to fight to the last.' Finally, they were down to fifty men but their task had been accomplished and they had enabled the rest of their comrades to rush up troops and check the German attempts to cross the Aisne.

Although the heroic stand of the Devons slowed down the German advance, it was not sufficient, on its own, to stop the great wave which edged its way forward until 8 August. On that day, with the Germans almost within sight of Paris, the Allied counter-attacks started the fight back. August proved a disastrous month for the Germans and from then on the Allied advance moved forward relentlessly until Germany asked for an armistice, which was granted on 11 November 1918 (at the eleventh hour of the eleventh day of the eleventh month).

BATTLEFIELD HELPERS

During World War I the Royal Flying Corps made a vital contribution in various aspects of warfare. It conducted important reconnaissance, it bombed and machine-gunned the enemy, and it fought aerial duels with its German counterparts. One of the most successful fighter aces was Captain Albert Ball, VC, a native of Nottingham. The Royal Flying Corps made astonishing progress during the war and played a decisive part in the final battles of 1918. In that year it was amalgamated with the Royal Naval Air Service to become the Royal Air Force. There are many exhibits in different museums and collections which relate to the Royal Flying Corps, some of them in RAF museums, but many in the Imperial War Museum.

The part played by technical and support arms is not usually given the credit it deserves. During World War I the contribution of the Royal Engineers in bridge and road construction and demolition, as well as in the Signals branch (which in 1920 would become the Royal Corps of Signals) are less well known than they deserve to be. One of the most glamorous, but most dangerous, activities of the war was despatch riding, but few of the old motor-cycles survived. The Army Service Corps, which jokingly referred to itself as 'Fred Karno's Army', after a music hall comedian of the day, had the onerous task of delivering ammunition and petrol up into the front line; the nature of their tasks meant that they were almost always exposed to enemy fire without much opportunity to hit back.

Wyvern Barracks, Exeter; Dorset Military Museum, Dorchester. Like a number of other regiments, the Devons are represented in more than one museum; the one in Wyvern Barracks includes material about the regiment from 1683 to 1958, and the Dorset Military Museum includes material from its amalgamation with the Dorsets, to become the Devon and Dorset Regiment, in 1958.

Castle Museum, Nottingham. Captain Ball's memorabilia and history are on display in the Castle Museum.

Royal Engineers Museum, Chatham; Royal Corps of Signals Museum, Blandford. These two museums tell the story of the Royal Engineers in World War I. Army signals were a Royal Engineers' responsibility pre-1920, and Royal Signals' thereafter.

Buller Barracks, Aldershot; Museum of Army Transport, Beverley. The Army Service Corps are now renamed the Royal Corps of Transport; they are excellently represented at Buller Barracks, and in the Museum of Army Transport.

Royal Army Veterinary Corps, Aldershot. Horses were employed by the thousand and displayed astonishing calm and patience under fire. The part played by the horse in warfare is shown in the various cavalry museums listed in the Gazetteer.

Imperial War Museum, London. The Imperial War Museum has now created an authentic World War I trench, complete with dark, noise and smells.

CHAPTER 10

WORLD WAR II

In the distant past the fate of a nation could be settled by a battle which was all over in one day, and sometimes within a few hours. Nowadays battles continue for weeks, and wars for years: World War I lasted for four years, World War II for six years. World War II saw the creation of new styles of fighting and also of new regiments. However, many of the so-called innovations of 1939–45 were not so much new inventions as more sophisticated applications of methods and weapons which had been used before.

(Opposite page) The World War II Cabinet War Rooms, Great George Street, London, which are open to the public.

EUROPE AND THE MIDDLE EAST

World War II began in September 1939 with the German invasion of Poland, an unprovoked attack which caused other countries, such

A 1932 Austin 7, reliable and sturdy transport. This is at the Royal Signals Museum.

as Britain and the great Commonwealth countries of Canada, Australia, New Zealand and South Africa, to declare war on Germany. Britain's declaration of war also brought in the countries of her empire, such as India, Ceylon, Burma, Malaya and many others. France's declaration brought in the countries of her own empire, which extended over one-third of Africa and included Indo-China as well. Russia was not originally concerned, for just before the war she had concluded a pact with her arch-enemy Hitler and his Nazi Germany. However, Russia's neutrality did not last for long, for first she invaded and occupied one half of Poland while the Germans were attacking the other, and two years later she herself was attacked and invaded by the Germans.

During the first six months of the war, Britain sent a small expeditionary force to France and assumed that after his Polish success Hitler would not risk a wider war. They found that this was far from true when Hitler sent his troops to conquer Norway and Denmark in April 1940, and a month later invaded Belgium, Holland, Luxembourg and France.

The German victory in Poland had already shown that the Germans were using what seemed to be a new form of warfare, and the invasion of France confirmed it. The new method, which Britain

The Spitfire seen in the foreground here played a vital part in the air defence of Britain. Behind is a German Focke-Wulf and in the centre an American P51 Mustang. These can be seen at the Imperial War Museum, London.

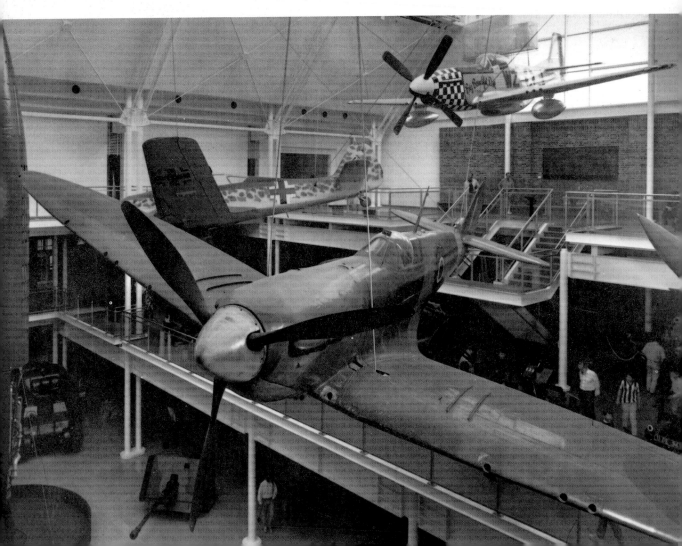

An exhibit of the 'Blitz Experience' at the Imperial War Museum, London. Blitz means lightning, and came to mean bombing, but was originally used in 'Blitzkrieg', or lightning war.

herself was also soon to employ, was to use tanks not merely as infantry support but to organise them in armoured divisions and to rely on them to use surprise, speed and mobility to reach their objectives. Previously, most attackers had used a preliminary artillery barrage, which removed all possibility of taking the enemy by surprise; now this was replaced by low-flying aircraft, dive-bombers and fighters 'shooting up' ground troops. In addition, the Germans used parachute troops to seize key areas ahead of the main attack then relied on tanks and infantry to support them.

These tactics caught the Belgians, Dutch, French and British by surprise and, as a result, the British were pushed out of France, mostly from Dunkirk, soon after which the French surrendered. During the fighting many British regiments had distinguished

A typical home in the 1939–1945 war. The father is away in the army, the mother is here at dusk, drawing the thick blackout curtains through which no chink of light could show. This exhibit can be seen at the Royal Engineers' Museum, Chatham.

Battle of Britain Museum, Hendon. See the Battle of Britain Museum for displays and memorabilia from this dramatic battle.

Cabinet War Rooms, Great George Street, London. The Cabinet War Rooms are now open.

themselves by their dogged resistance. As would be expected, the Guards, the Rifle Brigade (particularly at Calais) and the Seaforths, to name but a few, were among them. A British attempt to help Norway had also seen some courageous actions.

When the fighting in France ceased on 22 June, Britain took stock of a very perilous position. The army had lost all the valuable equipment it had sent to France and had had many of its most experienced soldiers killed or taken prisoner. Britain was now in great danger of invasion.

Churchill rallied the country with inspiring but grim speeches. He promised nothing but 'blood, toil, tears and sweat', but he struck a note of realism which people needed and wanted to hear. He told his chiefs-of-staff that they must 'set Europe ablaze', and for this must create 'commando regiments' which could land at night on the coast of occupied Europe and destroy the German garrisons. The story of the army Commandos is told best by Brigadier Peter Young, who was one of them, in his book *Storm from the Sea*. It describes the early raids and then the major battles like Dieppe, Sicily, and Normandy, and the campaigns in Burma.  The army Commandos, in conjunction with other 'special forces', are now setting up their museum at Inverary, where much of their training took place.

While the Commandos were getting into their stride, airborne forces were being created.  At the same time two other very different, but also very important, units were coming into existence. One was 'Special Operations Executive' (SOE) which planted agents (landed by parachute, submarine or canoe) all over Europe, where they worked behind the lines in conditions of great danger; the other was the Home Guard which, when Britain was in danger of being invaded in 1940, enrolled volunteers from people who were too old or too young to fight in the regular forces. In order to assist them and the regular forces, blockhouses, 'pill boxes' and camouflaged guns were installed. Many of these concrete blockhouses still stand today and are very obvious now that their camouflage has been removed, but during the war they were disguised with astonishing ingenuity to look like haystacks or shops or harmless buildings. Apart from these relics of Britain's hour of great peril (very much like that in the Napoleonic Wars nearly one-hundred-and-fifty years earlier), memorabilia will be found in local museums.

After the fall of France there was a desperate air battle for the control of the skies over the Channel. It lasted until late September 1940, but this story belongs to the RAF and not the army. However, it should not be forgotten that anti-aircraft guns, operated by men and women from the Royal Artillery and Auxiliary Territorial Service, played a significant part.

In 1941 Britain's military activities became centred on the Middle East, where initially her opponent was Italy, brought into

Army Commandos Museum, Inverary. The army Commandos, in conjunction with other 'special forces', are now setting up their museum at Inverary, where much of their training took place.

Airborne Forces Museum, Browning Barracks, Aldershot. The parachute regiments' history and equipment may be seen in the magnificent Airborne Forces Museum at Browning Barracks.

Spitfires on patrol, looking for enemy aircraft.

Two octagonal pillboxes: the one on the left is concealed in a haystack, the one on the right is camouflaged to conceal its outline.

the war because Mussolini, the Italian Fascist dictator, was Hitler's ally. Although the small British army in the Middle East was greatly outnumbered by the Italians, it soon proved that it was more than a match for them. The defeats it inflicted on the Italian armies convinced Hitler that he must support his Italian ally and turn the tide of the conflict. He therefore sent Erwin Rommel, a gifted tank general with an armoured corps (the Afrika Korps) to North Africa.

For the next two years a new form of warfare was to be seen in the Middle East. The 'Western Desert', as it was called, is an area 3,000 miles (4,800km) long by 1,500 miles (2,400km) deep and in the northern part of it many desperate battles were fought. The Germans had the advantage of far superior tanks and equipment, for they had been developing their tanks for years before the war. Their armoured regiments became known as 'Panzers', which was a shortened form of *Panzerkampfwagen*, meaning 'armoured fighting vehicle'. Added to this was the fact that Rommel was a genius at desert warfare and used his superior artillery, especially the notorious 88mm guns, with great skill. In July 1942 he reached what was known as 'the Alamein line', a defensive position which was the last obstacle to be overcome if he was to break through into Egypt, Alexandria, Cairo and Suez. Here, General Claude Auchinleck, the British Commander-in-Chief, took over tactical command of the battle and checked him.

At this moment, when Auchinleck should have received the highest praise, he was replaced by Churchill, who did not understand that the desperate British position was due to inferior tanks, artillery and aircraft, not only in quality but also in numbers. Auchinleck's replacement was General Bernard Montgomery, who was soon given enough tanks, guns and aircraft to give him a two to one superiority in the next large battle, which was also fought at Alamein. Rommel was again making a desperate attempt to break through before it was too late, but he did not realise he had such great odds against him, not the least of which were the three hundred Sherman tanks, a British design manufactured in America, which the Americans, who had now been drawn into the war, sent at the personal initiative of their President, Franklin Roosevelt.

Royal Armoured Corps Tank Museum and Royal Tank Regiment Museum, Bovington, Dorset. Visitors to the Tank Museum at Bovington will be able to inspect the Sherman tanks held there, as well as the 88mm guns and all the other weaponry of the Desert War.

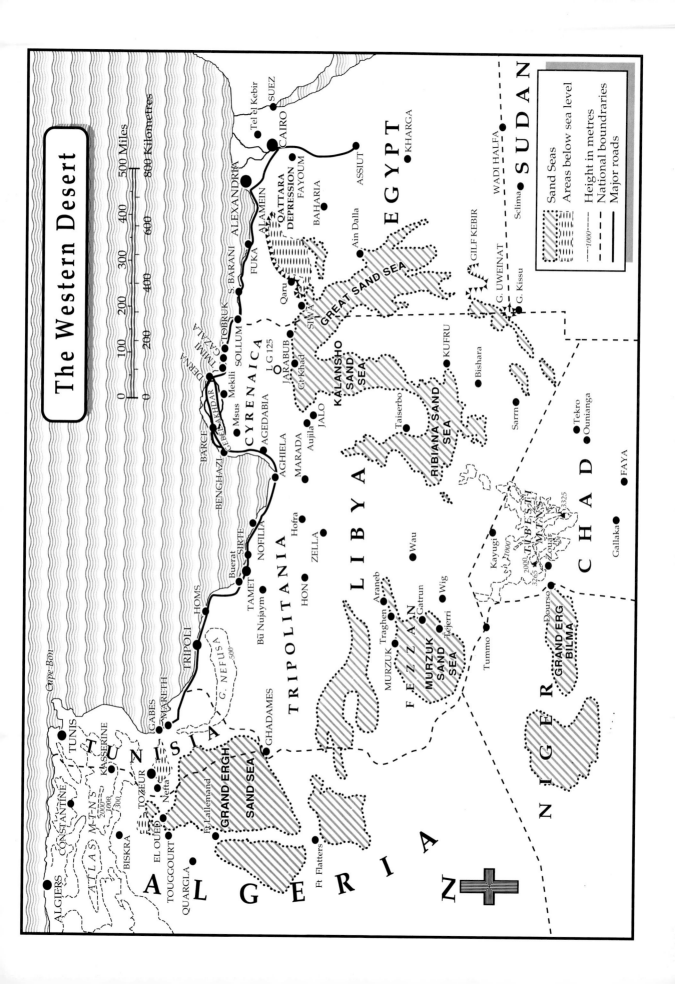

The Western Desert

A corner of the Imperial War Museum, London.

(Opposite above) A British Crusader tank in 1942.

(Opposite below) A Gloster Gladiator.

Churchill rides (uncomfortably) in a British Covenanter tank in 1942.

A British Cruiser tank in 1940.

Field Marshal Viscount Montgomery.

One of the most remarkable products of the desert war was the Special Air Service, which was the creation of David Stirling, a junior officer in the Scots Guards. He reasoned that as the German aircraft were too fast for the British to shoot them down in 1941, the best way to destroy them was to creep up to the German airfields, far behind their line, at night, and to blow them up with plastic explosive. General Auchinleck gave his permission to form a unit for this purpose. Stirling's idea had been to parachute men on to a nearby spot, but a disastrous attempt, ruined by a desert storm, convinced him that it would be more effective to approach on the ground using trucks (or later, jeeps). To do this, he enlisted the co-operation of the Long Range Desert Group, a regiment commanded by former desert explorers who had mastered the art of survival and navigation in the desert, but which was primarily a reconnaissance unit. In their new role the SAS destroyed three hundred and fifty German aircraft on the ground. The SAS's most famous soldier was Lt Col Blair Mayne, who won four DSOs, the *Croix de Guerre*, and the Legion of Honour. He was also a rugby international. After the end of the African campaign, the SAS went on to fight in Sicily and Italy, France and Germany, specialising in blowing up railways, ammunition dumps and enemy headquarters. Unfortunately, although the SAS could have a wonderful museum if it could put on display mementos of many of its daring deeds, it has now become so heavily involved in security that it is not possible, at the moment, to allow the public access to its barracks.

(Above) The Normandy beachhead
after the landing in June 1944.

(Below) A Sherman tank at the
D-day museum, Portsmouth.

After North Africa had been cleared, and 250,000 Germans and Italians taken prisoner, the Allies invaded Sicily. From Sicily they began a long campaign in Italy which only ended in 1945. By that time, the Allies, mainly Britain and America, had launched the invasion into France over the Normandy beaches and had fought their way across Europe until Germany surrendered. The invasion which took place on D-Day is displayed on a tapestry in the museum at Southsea and there are many other relics of D-Day in museums throughout Britain. Among those taking part was the Glider Pilot Regiment (which preceded the landings). Gliders were cheaper than aircraft and could land silently on targets far behind the enemy lines. They would also be used later at Arnhem in 1944, where they encountered disaster.

A DUKW amphibious landing vehicle as used on D-Day. This can be seen at the Museum of Army Transport, Beverley, Humberside.

(Opposite above) An exhibit at the D-Day Museum, Southsea.
(Below) exhibit in the D-Day Museum, Southsea.

THE FAR EAST

While the fighting continued in the Middle East and Europe, another, very different type of war was taking place in the Far East. This had begun in December 1941 with Japan's attack on Pearl Harbor (which brought America into the war) and which coincided

Army Air Corps Museum, Middle Wallop, Wiltshire.
The Glider Pilot Regiment is now incorporated in the Army Air Corps Museum at Middle Wallop.

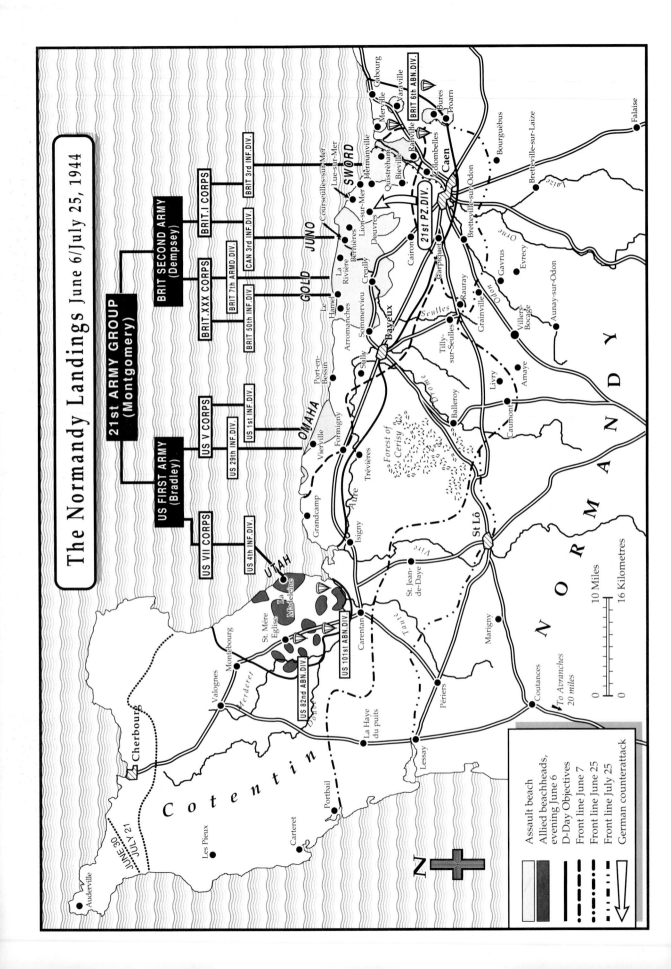

The Normandy Landings June 6/July 25, 1944

21st ARMY GROUP (Montgomery)

US FIRST ARMY (Bradley)

- US VII CORPS
  - US 4th INF. DIV.
- US V CORPS
  - US 29th INF. DIV.
  - US 1st INF. DIV.

BRIT SECOND ARMY (Dempsey)

- BRIT. XXX CORPS
  - BRIT 7th ARMD. DIV.
  - BRIT 50th INF. DIV.
- BRIT. I CORPS
  - CAN 3rd INF. DIV.
  - BRIT 3rd INF. DIV.

US 82nd ABN. DIV.
US 101st ABN. DIV.

BRIT 6th ABN. DIV.

21st PZ. DIV.

UTAH
OMAHA
GOLD
JUNO
SWORD

Cherbourg
Auderville
Les Pieux
Carteret
Portbail
La Haye du puits
Lessay
Montebourg
Valognes
Ste Mère Eglise
La Madeleine
Carentan
Périers
Coutances
Marigny
St. Jean-de-Daye
St Lô
Balleroy
Caumont
Grandcamp
Isigny
Trévières
Formigny
Vierville
Port-en-Bessin
Le Hamel
Arromanches
Sommervieu
Sully
Bayeux
Tilly-sur-Seulles
Livry
Amaye
Aunay-sur-Odon
Villers-Bocage
Grainville
Rauray
Gavrus
Evrecy
Garpiquet
Bretteville-sur-Odon
Cairon
La Rivière
Creully
Bernières
Courseilles-sur-Mer
Lue-sur-Mer
Douvres
Lion-sur-Mer
Hermanville
Bieville
Ouistreham
Colombelles
Caen
Ranville
Bures
Proarm
Varaville
Merville
Cabourg
Bretteville-sur-Laize
Bourguébus
Falaise

Montebourg
Valognes

JUNE 30
JULY 21

To Avranches
20 miles

Cotentin

N O R M A N D Y

Forest of Cerisy

Merderet
Douve
Taute
Vire
Aure
Drôme
Seulles
Odon
Orne
Laize

N

Legend

	Assault beach
	Allied beachheads, evening June 6
....	D-Day Objectives
–·–·–	Front line June 7
— — —	Front line June 25
–··–··–	Front line July 25
⟵	German counterattack

10 Miles
16 Kilometres
0
0

Mule transport in the Italian campaign.

with other surprise attacks on Hong Kong and Malaya. The Japanese, who had been seriously underrated as a fighting force, now showed that they had numerous warships, excellent modern aircraft and tanks, and millions of fanatically inspired soldiers. In Hong Kong the Royal Scots and the Middlesex put up a stout defence but were soon overwhelmed. In Malaya, where the Japanese had complete air and naval superiority, several units fought desperate rearguard actions before the inevitable fall of Singapore. Among those who fought doggedly in this campaign were the Leicesters and the East Surreys, the 2nd Argylls and a number of Royal Artillery batteries. The Japanese also invaded Burma, but were finally brought to a halt on the frontiers of India. Soon there followed epic battles at Kohima and Imphal. When the Japanese tried to break through Kohima in 1944, the opposing armies were at one period separated by the width of a tennis court (literally). The Royal West Kents (in which Lance Corporal Harman won a VC), the Northamptons, the Durham Light Infantry and the Dorsets were but a few of the regiments which added to their regimental honours.

The campaign to recapture Burma was particularly arduous

German one-man submarine.

Field Marshal Viscount Slim, mastermind of the Burma campaign.

because it was fought in a jungle which provided its own set of enemies: stifling heat, disease, insects, leeches, torrential rain and steep, slippery hills; the Japanese were a stubborn and resourceful enemy who made full use of jungle cover. In the earlier stages of the fight to recover Burma, Britain had sent in two 'Long Range Penetration Groups', which became known as the 'Chindits' (after a mythical lion). The Chindit expedition was a great boost to the morale of the Allied forces, who proved that the Japanese were not the invincible jungle fighters they had been thought to be; however, the strain of the Allies' deep incursion into the jungle, where the wounded had to be left behind to their fate, took an enormous toll on the survivors.

(Overleaf) American P51s were very effective in the Pacific. This one is at the Imperial War Museum, London.

After the Burma campaign was completed and the British army was preparing to invade Malaya, an American force (which had been approaching Japan by a series of hard-fought battles from island to island in the Pacific) dropped two atomic bombs, one on Hiroshima and the other on Nagasaki. Although the casualties caused by these bombs were less than had been inflicted by earlier conventional bombing raids, the effect of the two nuclear explosions was sufficient to shock the Japanese, already at the point of defeat, into surrender. Subsequently, much publicity was given to the effects of the bombs, which many said were inhumane, but none at all to the hideous cruelties inflicted by the Japanese on the prisoners they had taken years earlier, nor to the atrocities their armies had committed on civilians in the countries they had overrun. Unlike the Germans, who have a strong sense of guilt for atrocities committed by the troops and concentration camp staff during the war, the Japanese have never expressed the slightest regret for their behaviour against prisoners and non-combatants, preferring to pretend that it never happened.

Imperial War Museum, London. The Imperial War Museum now has a section which shows what the fate of a prisoner-of-war of the Japanese was likely to be, with other 'souvenirs' from survivors of the Burma–Thailand railway (which was constructed with a death for every sleeper) and other examples of Japanese slave-labour.

UNITS OF SPECIAL SERVICE

The part played by units other than infantry during the war has tended to be obscured. The Royal Engineers developed an astonishing speed at bridge construction, of which the famous 'Bailey' bridges were an outstanding example. Sir Donald Bailey was a 'bof-

'Gazelle', the smallest locomotive ever to operate on the standard-gauge railways of the UK, at the Museum of Army Transport, Beverley.

A display at the Tank Museum, Bovington, Dorset.

fin' (the word applied to brilliant 'back room' inventors), who used his engineering training to design a strong but light bridge which could be constructed in sections; the Royal Engineers made the fullest use of it.

The Royal Army Ordnance Corps ⌐ᵒ⌐ played a vital part in recovering and repairing wrecked tanks; some of them had been abandoned by the enemy but, after repair, were converted to use by the British. Neither the RAOC nor its offshoot, the Royal Electrical and Mechanical Engineers, ⌐ᵒ⌐ are normally thought of when people are considering feats of great initiative and daring, but the courage required to venture into No Man's Land in the dark and to recover a tank or gun which might have been booby-trapped by the enemy for just such a contingency, was of the highest order. To the RAOC and REME, however, it was just part of the job. During the Western Desert phase of the war, the speed at which broken-down vehicles could be recovered, repaired and put back into action was a vital factor in the success of the campaign. The other task which they handled with astonishing coolness was defusing unexploded enemy bombs.

Every unit, large or small, wherever it was, was linked to others by the Royal Corps of Signals. ⌐ᵒ⌐ They established communication by line, despatch riders and radio, and the hazards of being a lineman, called out to repair lines broken by enemy shelling, were

Royal Army Ordnance Corps Museum, Deepcut, near Camberley, Surrey. Contains regimental relics and displays.

Royal Electrical and Mechanical Engineers Museum, Arborfield, Reading. Here you will find a full display of regimental memorabilia.

Royal Corps of Signals Museum, Blandford Camp, Dorset. The Royal Corps of Signals, have a magnificent museum at Blandford Camp. Here you will find an example of an Enigma machine.

The German Enigma machine, which made 'unbreakable' codes, but we 'cracked' them.

many. Linked to the Royal Signals was 'Phantom', a regiment which, by liaison and by listening in the forward areas, obtained vital urgent information which was immediately sent back to GHQ. Phantom does not have a museum, but material relating to it may be found in other museums, for it consisted not only of Royal Signals but members of many other regiments, too.

The greatest contribution of the Royal Signals, however, was undoubtedly the part it played in the 'Y' service. 'Y' was the name given to the organisation (shared between the Royal Navy, the Army and the RAF) which monitored all enemy broadcasts. It required highly skilled operators, with a knowledge of foreign languages, who would listen with total concentration to transmissions of every variety, in cipher, code or plain language, and duly record

A searchlight unit at the Royal Engineers' Museum, Chatham.

them. This work entailed the operators living for many years in isolated, lonely and dangerous places, forbidden to tell their family or friends about their occupation or where they were; consequently, many operators whose lives and health were daily at risk, were suspected by their relatives of having an easy safe job in a backwater far removed from the conflict. Most of their 'intercepts' were decoded by the Enigma machine and passed to the Intelligence analysis centre at Bletchley, England. 'Enigma' machines were ingenious instruments, resembling typewriters, which could instantly encipher or decipher messages with several million possible permutations. The whole process was code-named 'Ultra' or, when it was performed against the Japanese in the Pacific, 'Purple'. Ultra did not mean that the Allies could decode *all* the enemy transmissions, but it enabled them to unravel enough to frustrate many enemy plans. Interception was linked with Deception, by which the Allies persuaded the Germans to make the wrong

Sherman tanks during an interlude in the battle for Germany 1945.

A suitcase radio, used by SOE and other intelligence agents working in enemy territory. This is at the Royal Signals Museum, Blandford.

Montgomery's staff car, a Rolls Royce 'Wraith' used in the North-West Europe campaign.

moves. Successful deception was practised to deceive the Germans about troop deployment at Alamein, the installation of dummy guns and airfields, and the landing areas in North Africa and Normandy (which Hitler thought would be around Calais). Constructing camouflaged or deception sites was the province of Royal Engineers, but the concepts were the concern of the Intelligence Corps, who have a museum at Ashford (Templer Barracks) in Kent.

Many units which played a vital part in World War II have now disappeared or been merged with others. They include many county yeomanry regiments which became artillery, tank units or signals. Their history may be seen in corners of larger museums, but this is no reflection on the value of their contribution to victory.

One of the many aircraft on display at Duxford, Cambridgeshire.

Royal Army Education Corps Museum, Beaconsfield; Royal Army Chaplains Museum, Bagshot; Royal Army Medical Corps Museum, Ash Vale, Aldershot. These are all excellent museums with many World War II exhibits.

CHAPTER 11

WARS SINCE 1945

When the guns stopped firing in 1945 it was hoped that it would bring 'peace in our time', to quote the words of Neville Chamberlain after he returned from his futile visit to Hitler at Munich in 1938. However, it was soon found that 'the price of peace is eternal vigilance' and the British army, although it was much reduced in size, prepared for its post-war peace-keeping role after 1945.

KOREA

Only five years passed before the British army was engaged in what was called a 'limited' war, the word 'limited' meaning that conventional, rather than nuclear, weapons would be used. At the end of World War II, Korea, which was to all intents and purposes a Japanese colony, was liberated by the Russians approaching from the north and the Allies coming up from the south. Commanders agreed that each army would halt its forces on the 38th parallel. During the next five years the south followed the path to democracy, while the north, which was firmly under Russian control, became a Communist state. This happened at the height of the 'Cold War', when Stalin and his Kremlin associates were ready to take advantage of any form of weakness in a neighbouring state. South Korea looked particularly vulnerable as her giant neighbour China had already established a Communist regime. In June 1950 the North Koreans suddenly crossed the 38th parallel and invaded the South. The United Nations was taken by surprise but soon rallied and despatched American and British troops. These, under the command of General MacArthur, proved more than a match for the North Koreans, but MacArthur's progress alarmed the Chinese, who, in consequence, came in huge numbers to support the North Koreans. The campaign then moved back into South Korea but the Chinese were eventually checked and the conflict continued with trench warfare of a type similar to that of World War I. By this time several British regiments had distinguished themselves, particularly the Middlesex and the Argylls. The final stages of the fighting, before the peace talks began, saw the Duke of Wellington's Regiment 🔫 engaged in relentless battles to retain possession of 'The Hook', a piece of valuable tactical ground which jutted out into the Chinese lines. Needless to say, many other British regiments and corps won battle honours in this brief but hard-fought war.

Duke of Wellington's Regiment, Bankfield Museum, Halifax. Here you will see exhibits relating to the regiment.

THE MIDDLE AND FAR EAST

All this time trouble was brewing in Malaya, Egypt, Aden, Cyprus and East Africa. Developments in those areas required regiments to operate in an anti-terrorist role. The British army's first post-war experience of what became called 'internal security' (IS) and, when it was more active, 'counter revolutionary warfare' (CRW) had come in Palestine in 1948, just before that country became the independent state of Israel. 'Peace keeping' there had been the prevention of conflict between Arabs and Jews, but the British army's attempts to preserve law and order led to its being the target of both sides' animosity.

In Malaya the anti-terrorist role was much larger. A strong Communist group, many of whose members had been trained and armed as anti-Japanese guerrillas, now decided that it would take over control of Malaya by murdering and intimidating workers on plantations and in tin mines. It was easy to do this from camps deep in the jungle and, when the British army was brought in to help eradicate this threat to the future of a democratic Malaya, it had a difficult task tracking down and killing or capturing the very elusive and skilled Communist terrorists. One of the earliest regiments to distinguish itself in this field was the Suffolks. Counter-terrorist operations against opponents using hit-and-run tactics from well-hidden jungle bases required a new technique of warfare. The Suffolks adapted to it quickly, even though they had many (originally) inexperienced national servicemen in their ranks. Part of the task was to win the 'hearts and minds campaign' – that is, to obtain the trust and co-operation of the local people.

> **Suffolk Regimental Museum, Gibraltar Barracks, Bury St Edmunds.** This musuem holds many regimental relics and displays.

The SAS were to bring deep penetration tactics to a fine art. Working in four-man groups, they tracked the terrorists to their remote camps and surprised them. The techniques of jungle survival which the SAS developed had a lasting effect on the clothing, equipment and weapons which would be used for these enterprises in the future. Among their innovations was parachuting into trees, but few wished to follow their example. Their jungle survival techniques were used by themselves and others in the mid-1960s when the Indonesians tried to invade Borneo and were soon outfought.

Further trouble came from Egypt in the 1950s when Abdul Nasser, head of the Egyptian Republic, seized the assets of the Suez Canal Company which was mainly owned by Britain and France. An Anglo-French invasion provoked strong reaction in many other countries; the United Nations was brought in and the invasion forces withdrew. Nasser was thus encouraged to stir up trouble throughout the Middle East, notably in Aden. This was a time when Britain was withdrawing from her imperial commitments as fast as possible. The British army spent time and blood in preserving law and order in Aden and the adjoining protectorates, but withdrew in 1967, leaving it to become the republic of South Yemen. In the days of the British Empire, Aden had been of great importance as a coaling station, but it is far from prosperous today.

There was a happier outcome to an operation which took place in Oman, on the Persian Gulf, in 1959. A rebel chieftain threatened the friendly state of Oman, but a small British force spearheaded by the SAS dislodged him from his mountain fortress.

A Beverley transport aircraft.

A 'Rhino', the ultimate cross-country
vehicle; however, only one was ever
built. It is at the Royal Transport
Museum, Beverley.

A Matilda tank.

Italian Human Torpedo craft.

Cyprus provided a different sort of problem. There, independence was hindered by the inability of the two groups involved, the Turks and the Greeks, to agree. The eventual outcome was partition and a permanent United Nations peace-keeping force of which Britain provides a notable component.

THE FALKLANDS

Apart from continuing internal security duties in Northern Ireland, the British army had a relatively peaceful period after 1970, but in 1982 was suddenly called upon to take part in an operation 8,000 miles (13,000km) away in a climate which is said to be the worst in the world. This was the Falklands War.

The Falkland Islands, on the edge of the Antarctic, had been a British dependent territory since 1833. All their eighteen hundred inhabitants were of British stock and wished to remain British. However, in April 1982, the Argentine government, at that time a brutal military dictatorship, decided to launch an unannounced attack and to capture the islands. The token garrison in the Falklands was soon overwhelmed but the immediate British response was to send a task force to recapture the islands. This was accomplished after vigorous fighting by Royal Marines, the Parachute Regiment, Scots and Welsh Guards, Gurkhas and others. The SAS took its usual unobtrusive but effective part. The Royal Signals brought off a remarkable feat by establishing communication with Britain via satellite within twenty minutes of landing.

In the RAMC museum at Ash Vale, Aldershot, is the operating table from a field hospital which was hit by an Argentinian bomb during the Falklands War. On the table was an Argentinian soldier whose life was being saved by the RAMC. He survived. The RAMC gives help professionally and impartially to friend or enemy in saving lives.

THE GULF WAR

On 13 August 1990 the Iraqui army invaded its neighbouring country of Kuwait. Saddam Hussein, the Iraqui leader, refused to withdraw despite strong diplomatic and economic pressures, and under the umbrella of the United Nations hundreds of thousands of Allied soldiers massed on the Kuwaiti/Saudi Arabian border. The Iraqui army, equipped with modern weapons, was believed to be highly trained and around a million strong.

During the closing months of 1990 the Allies made extensive preparations in Saudi Arabia. Many British forces were transferred from Germany at short notice, a feat requiring considerable logistic skill. As this was a full-scale campaign, every arm and service in the British army took some part in it. During the pre-battle preparations, units of the Royal Signals took up positions near the enemy lines in order to monitor, jam and dislocate enemy transmissions. Special Forces such as the SAS made deep forays into Iraqui territory, identifying targets, observing enemy dispositions, and

Royal Marines Museum, Eastney, Portsmouth. A full display of regimental relics.

planting various sabotage devices. Royal Engineers, and Royal Army Ordnance Corps, defused Iraqui-laid mines and engaged in demolitions.

The war finally started on 18 January 1991, the day the UN deadline for the removal of Iraqui troops ended. When fighting began, the public at home heard names which recalled World War II, particularly the North African campaign – among them, the 17th/21st Lancers, the Queen's Royal, Irish Hussars, the Royal Scots Dragoon Guards (the Greys), and the 14th/20th Hussars. These regiments had often fought side by side in battle. In this war their weapons were very different from those of earlier wars – horses had been replaced by Challenger tanks and Scorpion armoured cars. However, the essential spirit (and much of the daily conversation) was the same. The Royal Artillery also had a variety of new weapons, including the devastating Multiple Launch Rocket System. The Royal Corps of Transport, formerly the Royal Army Service Corps, performed miracles under near-impossible conditions. In the meantime the 'Desert Rats', as the 7th Armoured Brigade are proud to be called, were moving along the ground whilst the RAF swept the skies with Tornado bombers.

This war was one of great co-operation and co-ordination between armies of different countries, cultures and languages. In addition, it was a truly 'high-tech' war, one weapon battling to destroy the other. Casualties on the Allied side were light, although the Royal Fusiliers were unfortunate in having several of their soldiers killed accidentally. The battle was 'won' mainly in the sky, Allied bombers putting out of action most of the Iraqui artillery and weaponry. Saddam Hussein agreed to the United Nation demands for withdrawal on 28 February 1991, and offensive operations ceased.

For most of the soldiers, the Gulf War was often simply one against the poor conditions and boredom of life in the desert; for the pilot one of daily and frequent raids into enemy territory. Like the Korean War forty years earlier, this campaign to deter aggression established the British soldier as a member of an international army, working successfully under the banner of the United Nations.

THE MODERN SOLDIER

Today the British army maintains a peace-keeping rôle in many countries throughout the world. On ceremonial occasions it gives a display second to none. The British soldier has the ability to handle sophisticated weapons in his NATO rôle or to act in his 'fire-brigade' capacity by flying at short notice to any troublespot in the world. History shows that he is a formidable fighter in mountains or valleys, in jungle or desert, in Arctic or tropical climates. When not fighting he is kind and protective to those who seek his help. It has been rightly said, 'The British soldier is Britain's best ambassador.'

THE MODERN MUSEUM

Britain's military museums contain a fascinating blend of the old and the new, particularly in transport. Early armoured cars sit beside post-war Beverley aircraft. Tanks from 1917 are kept under the same roof as post-war 'Centurion' tanks. Nearby is, perhaps, a 'Ferret' Mk 5 'Swingfire' Guided Missile Vehicle.

Cavalry museums display historic guidons (pennants), but they also contain tanks. The achievements of the past are commemorated in a variety of ways and exhibits, which offer a unique chance to understand our heritage. Soldiers no longer march into battle to the skirl of bagpipes as they did two hundred years ago: in 1940 they marched to the sound of 'Matilda' Infantry tanks and heard the roar of aircraft overhead. When the Germans introduced modern rockets in 1944 they transformed warfare. Later would come the Polaris missile and similar weapons. Since World War II, museums have developed enormously and have now collected large quantities of exhibits from all areas of the soldier's life. Many long-kept secrets and prohibited places are now available to the general public.

British museums are not exclusively involved with British weapons. There are numerous captured 'trophies' on view, and an exhibit of special interest in the Imperial War Museum is the Italian Human Torpedo. In the later stages of the war, similar devices were used very effectively by the Allies.

In past centuries the army used to rely on the cavalry to break up enemy formations. Today it is done by aircraft such as the Tornado F16A which carries four missiles. However, whatever the period and the instrument, the courage and skill of the user is the all-important factor. That is the message which the modern museum conveys.

GAZETTEER OF MUSEUMS

Note: You are advised to check opening times with individual museums before visiting.

ENGLAND

BEDFORDSHIRE

Bedfordshire and Hertfordshire Regimental Museum, Luton Museum, Wardown Park, Luton
Tel: Luton (0582) 36941
Memorabilia of the Bedfordshire and Hertfordshire Regiment.

The Shuttleworth Collection, Old Warden Aerodrome, Biggleswade
Tel: Biggleswade (0767) 288
Comprehensive range of military aircraft and vehicles including a 1912 Blackburn monoplane, Sopwith Pup and 1941 Spitfire.

BERKSHIRE

94 (Berkshire Yeomanry) Signal Squadron Museum, TA Centre, Bolton Road, Windsor
Tel: Windsor (0753) 60600
Relics of this Signal Squadron's history.

The Household Cavalry Museum, Combermere Barracks, Windsor
Tel: Windsor (0753) 868222 ext 203
Excellent displays and dioramas.

REME Museum, Isaac Newton Road, Arborfield
Tel: Reading (0734) 760421 ext 2567
Engineering technology.

Windsor Castle, Windsor
Tel: Windsor (0753) 831118
Many military relics: the Norman keep gives an excellent idea of a Norman motte and bailey (mound and enclosure) castle.

BUCKINGHAMSHIRE

Royal Army Educational Corps Museum, Wilton Park, Beaconsfield
Tel: Beaconsfield (049 46) 6121 ext 286
Army education since 1815.

CAMBRIDGESHIRE

Ely Museum, Sacrist's Gate, High Street, Ely
Tel: Ely (0353) 2311
Memorabilia of the Cambridgeshire Regiment.

Imperial War Museum, Duxford
Tel: Cambridge (0223) 833963
Over 100 military aircraft as well as tanks, guns, etc, excellently displayed.

CHESHIRE

The Cheshire Military Museum, The Castle, Chester
Tel: Chester (0244) 27617
Memorabilia of the 5th Royal Inniskilling Dragoon Guards, 3rd Carabiniers, Cheshire Yeomanry and 22nd (Cheshire) Regiment.

The Grosvenor Museum, 27 Grosvenor Street, Chester
Tel: Chester (0244) 21616
Numerous displays of Chester's remarkable military history. (Chester was the base of the Roman XX Legion for 300 years.)

The Queen's Lancashire Regimental Museum, Peninsular Barracks, Warrington
Tel: Warrington (0925) 33563
Weapons of every variety; extensive archives.

CO DURHAM

Durham Light Infantry Museum and Art Centre, Aykley Heads, Durham
Tel: 091-384 2214
Items from the Durham Light Infantry, 68th Light Infantry, Durham Militia, volunteers and associated regiments; guns from World War II, Bren gun carrier.

CORNWALL

Duke of Cornwall's Light Infantry Regimental Museum, The Keep, Bodmin
Tel: Bodmin (0208) 2810
Items depicting this regiment's history.

Flambards Triple Theme Park, Clodgey Lane, Helston
Tel: Helston (0326) 573404
Helicopters, hovercraft, the Falklands War, 'Britain in the Blitz' exhibition.

CUMBRIA

The Border Regiment and King's Own Border Regiment Museum, Carlisle
Tel: Carlisle (0228) 32774
Situated in the 11th century castle which saw continuous fighting during the Border wars. Videos of various battles; diorama of Arnhem.

Westmorland and Cumberland County Museum, Penrith
Tel: Penrith (0768) 450
Mementoes of the yeomanry of the two counties which were joined in 1972.

DERBYSHIRE

Regimental Museum of the 9th/12th Royal Lancers, Derby City Museum and Art Gallery, The Strand, Derby
Tel: Derby (0332) 293111 ext 792
Also includes items from militia and volunteer regiments, the Sherwood Foresters and the Derbyshire Regiment.

DEVON

Cobbaton Combat Vehicles Museum, Littlehampton, Umberleigh, Nr Barnstaple
Tel: Chittlehamholt (076 94) 414
British and Canadian armoured fighting vehicles and equipment.

DORSET

Dorset Military Museum, The Keep, Dorchester
Tel: Dorchester (0305) 64066
Extensive displays, uniforms and weapons from the Dorset Regiment, Devon and Dorset Regiment, Dorset Militia and Volunteers and Queen's Own Dorset Yeomanry.

Royal Armoured Corps Tank Museum and Royal Tank Regiment Museum, Bovington Camp, Wareham
Tel: Bindon Abbey (0929) 462721 ext 463
Largest collection of armoured fighting vehicles in world; over 250 tanks. Theatres, simulators and exhibitions with cut-aways for interior inspection.

Royal Signals Museum, Blandford Camp, Nr Blandford Forum
Tel: Blandford (0258) 52581 ext 2248
Wide variety of equipment: motor bicycles, weapons, radios, early satellite dishes and an Enigma machine.

EAST SUSSEX

The Combined Services Museum, Redoubt Fortress, Royal Parade, Eastbourne
Tel: Eastbourne (0323) 35809
Exhibits of numerous regiments; models.

Langton House, Abbey Green, Battle
Tel: Mr Denny (Battle Historical Society) (04246) 4295
Diorama of the Battle of Hastings; relics.

Martello Tower 73: The Invasion Museum, The Wish Tower, King Edward's Parade, Eastbourne
Tel: Eastbourne (0323) 35809
Exhibits of invasion defences.

Military Heritage Museum, 1 Albion Street, Lewes
Tel: Lewes (0273) 3139
Interesting cavalry items.

ESSEX

Colchester and Essex Museum, The Hollytrees, High Street, Colchester
Tel: Colchester (0206) 712481
Items relating to the Essex Volunteer forces.

Essex Regiment Museum, Oaklands Park, Moulsham Street, Chelmsford
Tel: Chelmsford (0245) 260614
Numerous exhibits including captured trophies.

GLOUCESTERSHIRE

Museum of the Gloucestershire Regiment, 31 Commercial Road, Gloucester
Tel: Gloucester (0452) 22682
Museum of this distinguished regiment which wears the Back Cap Badge from the Battle of Alexandria (1801).

HAMPSHIRE

Airborne Forces Museum, Browning Barracks, Aldershot
Tel: Aldershot (0252) 24431 ext 4619
Superb displays including a Falklands Room.

Aldershot Military Museum, Queen's
Avenue, Aldershot
Tel: Aldershot (0252) 314598
1900 barrack room, Canadian Army displays,
dioramas, guns, tanks.

**HMS Alliance, Holland I and Royal Navy
Submarine Museum**, Haslar Jetty Road,
Gosport
Tel: Gosport (0705) 529217
Holland I was the Royal Navy's first
submarine, and was salvaged after 69 years
under water. Alliance is a restored 1945
submarine. Midget submarines.

Army Catering Corps Museum, St Omer
Barracks, Aldershot
Tel: Aldershot (0252) 24431 ext 2616
Closed temporarily.

Army Physical Training Corps Museum,
Corps Depot, Queen's Avenue, Aldershot
Tel: Aldershot (0252) 24431 ext 2131
An important aspect of army life fascinatingly
portrayed.

Museum of Army Flying, Army Air Corps
Centre, Middle Wallop
Tel: Andover (0264) 62121
History of army flying, aircraft, gliders,
helicopters.

The Peninsula Barracks, Winchester
Tel: Winchester (0962) 63846
The Barracks house four magnificent
museums: the Light Infantry, the Gurkhas,
the Royal Hussars and the Royal Green
Jackets (formerly the Rifle Brigade, 60th
Rifles and Oxford and Bucks Light Infantry).
Displays of the Battle of Waterloo, replica
Baker rifles, jungle warfare scenes, displays of
historic battles.

**Queen Alexandra's Royal Army Nursing
Corps Museum**, RHQ QARANC, Royal
Pavilion, Farnborough Road, Aldershot
Tel: Aldershot (0252) 24431 ext 4301
Army nursing since 1854.

Royal Army Dental Corps Museum,
RADC, Evelyn Woods Road, Aldershot
Tel: Aldershot (0252) 24431 ext 2782
Early dental instruments.

Royal Army Medical Corps Museum,
Keogh Barracks, Ash Vale, Nr Aldershot
Tel: Aldershot (0252) 24431 ext 5212
Relics of every campaign from 1600. Replica of
Captain Chavasse's unique VC and bar.

Royal Army Pay Corps Museum, Worthy
Down, Winchester
Tel: Winchester (0962) 880880 ext 2435
Interesting records.

Royal Army Veterinary Corps Museum
Temporarily closed pending new site.
Enquiries tel: Aldershot (0252) 24431 ext
3527.

Royal Corps of Transport Museum, Buller
Barracks, Aldershot
Tel: Aldershot (0252) 24431 ext 3834
Exhibits and dioramas from 1795 to present
day.

HEREFORD AND WORCESTER

**Hereford Regiment and Light Infantry
Museum**, TA Centre, Harold Street, Hereford
Tel: Hereford (0432) 272914
Interesting items from the Napoleonic period.

Worcestershire Regimental Museum,
Worcester City Museum, Foregate Street,
Worcester
Tel: Worcester (0905) 25371
Regimental, militia and volunteer relics.

HERTFORDSHIRE

Hertfordshire Regiment Museum, 18 Bull
Plain, Hertford
Tel: Hertford (0992) 582686
Regimental exhibits.

The Hitchin Museum, Paynes Park, Hitchin
Tel: Hitchin (0462) 34476
Yeomanry and artillery items of the
Hertfordshire Regiment.

HUMBERSIDE

Hull City Museum, Wilberforce and
Georgian Houses, High Street, Hull
Tel: Hull (0482) 222737
Items from the East Yorkshire Regiment,
militia and volunteers.

Museum of Army Transport, Flemingate, Beverley
Tel: Hull (0482) 860445
Comprehensive collection of all kinds of military vehicles including locomotives.

KENT

The Buffs Regimental Museum, Royal Museum, High Street, Canterbury
Tel: Canterbury (0227) 452747
The Buffs were the Royal East Kent Regiment whose long and distinguished history is exhibited here.

Hever Castle, Hever, Nr Tonbridge
Tel: Edenbridge (0732) 865224
Memorabilia of Kent and County of London Yeomanry (the Sharpshooters), a tank regiment in World War II.

Intelligence Corps Museum, Templar Barracks, Ashford
Tel: Ashford (0233) 625251 ext 208
Exhibits showing the history of intelligence gathering from Elizabethan times to the present day.

Quebec House, Quebec Square, Westerham
Tel: Westerham (0959) 62206
Many General Wolfe items and display of the Battle of Quebec.

The Queen's Own West Kent Regiment Museum, St Faith's Road, Maidstone
Tel: Maidstone (0622) 54497
Items from the history of the regiment, the West Kent Militia and the 20th London Regiment.

The Queen's Regimental Museum, Inner Bailey, Dover Castle, Dover
Tel: Dover (0304) 457411 ext 4253
The Buffs, the East Surreys, Royal West Kents, Royal Sussex and Middlesex Regiments.

Royal Engineers Museum, Prince Arthur Road, Gillingham
Tel: Medway (0634) 44555 ext 2312
The story of military engineering and its equipment throughout the centuries.

Walmer Castle, Deal
Tel: Deal (0304) 364288
Wellington relics including his narrow bed.

LANCASHIRE

Blackburn Museum and Art Gallery, East Lancashire Regimental Museum, Museum Street, Blackburn
Tel: Blackburn (0254) 667130
Trophies include a French drum captured at Waterloo and an eagle taken at Salamanca.

The British in India Museum, Sun Street, Colne
Tel: Colne (0282) 63129
Model soldiers, paintings, working model of the Kalka-Simla railway, diorama of the last stand of the 44th (Essex) Regiment at Gandamak, uniforms of the late Field Marshal Sir Claude Auchinleck.

County and Regimental Museum, Stanley Street, Preston
Tel: Preston (0772) 264075
Mementoes of Queen's Lancashire (Loyals and Lancashire) Regiment, Yeomanry and the 14th/20th King's Hussars. Napoleonic relics; authentic reconstruction of World War I trench.

Lancaster City Museum, Lancaster
Tel: Lancaster (0524) 64637
Memorabilia of King's Own Royal Regiment, militia, volunteers and the Abyssinia Expedition of 1868.

Regimental Museum of the Lancashire Fusiliers, Wellington Barracks, Bury
Tel: Bury 061–764 2208
History of the Lancashire Fusiliers, relics of General Wolfe and Napoleon.

Regimental Museum, Loyal Regiment (North Lancashire), Fulwood Barracks, Preston
Tel: Preston (0772) 716543 ext 2362
In 1759 the regiment was the 81st (Invalids). It has also been the Aberdeen Highlanders and the Loyal Lincolns – now the Queen's Lancashire Regiment.

Towneley Hall Art Gallery, Burnley
Tel: Burnley (0282) 24213
Exhibits of East Lancashire Regiment (3rd Battalion), recruited mainly around Burnley. Relics of General Scarlett and the Charge of the Heavy Brigade in the Crimean War in 1854.

LEICESTERSHIRE

Royal Leicestershire Regimental Museum, The Magazine, The Newarke, Oxford Street, Leicester
Tel: Leicester (0533) 555889
The 'Tigers', now incorporated in the Royal Anglian Regiment, have an interesting museum with many exhibits in the 14th century gatehouse.

Rutland County Museum, Catmos Street, Oakham
Tel: Oakham (0572) 365
Memorabilia of mainly volunteer regiments.

LINCOLNSHIRE

17th/21st Lancers Regimental Museum, Belvoir Castle, Nr Grantham
Tel: Grantham (0476) 67413 ext 252
Items from both the 17th and 21st, including the famous Omdurman charge.

Museum of Lincolnshire Life, Lincoln
Tel: Lincoln (0522) 528448
Memorabilia of the Royal Lincolnshire Regiment and the Lincolnshire Yeomanry.

LONDON

Guards Museum, Wellington Barracks, Birdcage Walk, London SW1
Tel: 071-930 4466 ext 3253
History of all the Foot Guards – Grenadier, Coldstream, Scots, Irish and Welsh – whose fighting record is second to none.

Hendon Royal Air Force Museum, London NW9
Tel: 081-205 2266
Aircraft, balloon units, Royal Flying Corps, Royal Naval Air Service, as well as the story of the RAF. Many foreign (enemy!) aircraft on display.

Imperial War Museum, Lambeth Road, London SE1
Tel: 071 416 5000
Refurbished national museum covering the history of military operations since 1914. Displays of every variety: World War I trench warfare, Japanese POWs in World War II, the Blitz. Rockets, aircraft, guns; excellent archives.

Middlesex Regimental Museum, Bruce Castle, Lordship Lane, Tottenham, London N17
Tel: 081–808 8772
Museum of the 'Diehards', the Middlesex Regiment, now incorporated with the Queen's.

Museum of Artillery, The Rotunda, Woolwich, London SE18
Tel: 081-854 2242 ext 3127
Guns of all kinds: 25-pdrs, 17-pdrs, screw guns, large guns and personal weapons. Earliest is a 1346 bombard (perhaps used at Crécy), latest are rockets.

National Army Museum, Royal Hospital Road, Chelsea, London SW3
Tel: 071-730 0717
Story of the British army from 1485 to the present day, also the Indian army before independence in 1947. Uniforms, weapons, numerous dioramas – even the skeleton of Napoleon's horse, 'Marengo'.

The Royal Armouries, The Tower of London, London EC3
Tel: 071-709 0765
National collection of arms and armour.

Royal Artillery Regimental Museum, Old Royal Military Academy, Academy Road, Woolwich, London SE18
Tel: 081-854 2242 ext 3128
Uniforms, history and campaigns of the Royal Artillery.

Royal Fusiliers Museum, The Tower of London, London EC3
Tel: 071-709 0765 ext 295
The Royal Fusiliers, formed in 1685, were merged with the other Fusilier regiments in 1968. Excellent dioramas of the Alma, Mons and Cassino.

Royal Hospital Chelsea Museum, Royal Hospital Road, Chelsea, London SW3
Tel: 071-730 0161
Relics of the Royal Hospital, fine chapel and dining hall, Wellington Hall.

Science Museum, Exhibition Road, South Kensington, London SW7
Tel: 071-938 3000
Many aeronautical items including space travel equipment.

Victoria and Albert Museum, Cromwell Road, South Kensington, London SW7
Tel: 071-589 6371
Numerous examples of armour.

Wallace Collection, Hertford House, Manchester Square, London W1
Tel: 071-935 9687
European and oriental arms and armour.

Wellington Museum, Apsley House, 149 Piccadilly, London W1
Tel: 071-499 5676
Apsley House was bought by the 1st Duke of Wellington in 1817 and presented to the nation by the 7th Duke in 1947. Paintings, decorations and relics of the 1st Duke.

MERSEYSIDE

Museum of the King's Regiment, Liverpool Museum, William Brown Street, Liverpool
Tel: 051-207 0001
The King's Regiment was formed when the King's Liverpool Regiment and the Manchester Regiment merged in 1958. The latter began as the 63rd (American) Foot; the King's Liverpool have also been the Queen's.

NORFOLK

Cholmondeley Collection of Model Soldiers, Houghton Hall, King's Lynn
Tel: (0485) 22569
20,000 model soldiers perfectly displayed.

The Royal Norfolk Regiment Museum, Britannia Barracks, Norwich
Tel: Norwich (0603) 628455
Large, comprehensive museum of a distinguished regiment.

NORTHAMPTONSHIRE

Museum of the Northamptonshire Regiment, Abington Park Museum, Northampton
Tel: Northampton (0604) 35412
The Northamptonshire and Lincolnshire Regiments are now merged and form part of the Royal Anglian.

Royal Pioneer Corps Museum, Simpson Barracks, Wootton, Northampton
Tel: Northampton (0604) 62742
The interesting but little known history of the Pioneer Corps.

NORTHUMBERLAND

5th Royal Northumberland Fusiliers Regimental Museum, The Abbots Tower, Alnwick Castle, Alnwick
Tel: Alnwick (0665) 602152
Memorabilia of the 5th Royal Northumberland Fusiliers.

Chesterholm Museum, Vindolanda, Bardon Mill, Hexham
Tel: Bardon Mill (049 84) 277
Excavated relics from Vindolanda fort, the base for Roman soldiers stationed along Hadrian's Wall.

Clayton Memorial Museum and Chester Roman Fort, Chollerford, Hexham
Tel: Humshaugh (043 481) 379
Items from Roman forts along Hadrian's Wall.

Hexham Roman Army Museum, Carvoran, Greenhead, Hexham
Tel: Gilsland (069 72) 485
Depicts the Roman soldier and his way of life.

King's Own Scottish Borderers Museum, The Barracks, Berwick-upon-Tweed
Tel: Berwick-upon-Tweed (0289) 307426
History of the regiment since 1689; exhibits and trophies.

NORTH YORKSHIRE

4th/7th Royal Dragoon Guards Museum, 3 Tower Street, York
Tel: York (0904) 642038
Excellent historical display.

Castle Museum, Tower Street, York
Tel: York (0904) 653611
Large, comprehensive museum; diorama of the Civil War, uniforms from local militia and yeomanry regiments. (World War I trench no longer on display.)

The Green Howards Museum, Trinity Church Square, Richmond
Tel: Richmond (0748) 2133
History from 1688 to the Falklands. 80 uniforms, 3,000 medals with VCs and GCs.

The Museum of the Prince of Wales Own Regiment, Yorkshire Museum, 3A Tower Street, York
Tel: York (0904) 642038
Museum of the former East Yorkshires and West Yorkshires; many VCs.

RAF Regiment Museum, RAF Regiment
Depot, Catterick
Tel: Richmond (0748) 811441 ext 202
History of the RAF Regiment, formed to
protect airfields from ground or parachute
attack. Anti-aircraft guns, missiles, vehicles.

NOTTINGHAMSHIRE

Newark Museum, Appleton Gate, Newark
Tel: Newark (0636) 702358
Relics of the Sherwood Foresters; TA
material.

The Sherwood Foresters Museum, The
Castle, Nottingham
Tel: Nottingham (0602) 785516
Exhibits and medals from the Sherwood
Foresters Regiment, now amalgamated with
the Worcestershire Regiment under the title
of the Worcesters and Foresters.

OXFORDSHIRE

**Regimental Museum of the Oxfordshire
and Buckinghamshire Light Infantry**, TA
Centre, Slade Park, Headington, Oxford
Tel: Oxford (0865) 778479
Mementoes of the 'Ox and Bucks', now part of
the Royal Green Jackets; also of the Queen's
Own Oxfordshire Hussars.

SHROPSHIRE

**1st Queen's Dragoon Guards Regimental
Museum**, Clive House, College Hill,
Shrewsbury
Tel: Shrewsbury (0743) 54811
Memorabilia of regiment that began as the
Queen's Horse in 1682, King's Dragoon
Guards in 1746 and amalgamated with the
Queen's Bays (2nd Dragoons) in 1959.
Interesting Waterloo material.

The Shropshire Regimental Museum, The
Castle, Shrewsbury
Tel: Shrewsbury (0743) 585516
Now part of the Light Infantry, the KSLI has a
great fighting tradition which is well
documented here.

SOMERSET

Fleet Air Arm Museum, Royal Naval Air
Service, Yeovilton, Ilchester
Tel: Ilchester (0935) 840565
Comprehensive display of historic aircraft.

Somerset Military Museum, County
Museum, The Castle, Taunton
Tel: Taunton (0823) 333434
History of the Somerset Light Infantry and
associated Yeomanry Regiment; very strong
on World Wars I and II.

**South Somerset District Council
Museum**, Hendford, Yeovil
Tel: Yeovil (0935) 24774
Weapons from 18th and 19th centuries.

SOUTH YORKSHIRE

Cannon Hall Museum and Art Gallery,
13th/18th Royal Hussars Regimental
Museum, Cannon Hall, Cawthorne, Barnsley
Tel: Barnsley (0226) 790270
Exhibits of the 13th/18th Royal Hussars.

Doncaster Museum and Art Gallery,
Chequer Road, Doncaster
Tel: Doncaster (0302) 734287
Items from the King's Own Yorkshire Light
Infantry, York and Lancaster Regiment,
Yorkshire Dragoons and various yeomanry
and militia regiments.

York and Lancaster Regimental Museum,
Central Library and Art Centre, Walker Place,
Rotherham
Tel: Rotherham (0709) 2429
Although the regiment is now in abeyance, its
history (1759–1968) is very well displayed.
Viewing by appointment only.

STAFFORDSHIRE

**16th/5th Lancers and the Staffordshire
Yeomanry Regimental Museum**, Kitchener
House, Lammascote Road, Stafford
Tel: Stafford (0785) 45840
Weapons, uniforms, videos.

Museum of the Staffordshire Regiment,
Whittington Barracks, Lichfield
Tel: Lichfield (0543) 433333
Numerous exhibits and weapons from many
campaigns; 8 VCs.

SUFFOLK

Norfolk and Suffolk Aviation Museum,
Flixton, Bungay
Tel: Thurton (050 843) 778
History of military aviation including 16
aircraft on display.

Suffolk Regimental Museum, The Keep,
Gibraltar Barracks, Out Risbygate Street,
Bury St Edmunds
Tel: Bury St Edmunds (0284) 769834
Mementoes, exhibits and historical data of the
Suffolk Regiment (now part of the Royal
Anglian); also material on the Cambridgeshire
Regiment (TA).

SURREY

Museum of Army Chaplains, Bagshot Park,
Nr Camberley
Tel: Camberley (0276) 71717
Exhibits include a remarkable VC. By
appointment only.

**National Army Museum (Indian Army
Exhibition),** RMA Sandhurst
Tel: Camberley (0276) 63344 ext 2457

**The Queen's Royal Surrey Regimental
Museum,** Clandon Park, West Clandon,
Guildford
Tel: Guildford (0483) 223419
Memorabilia of the Queen's Royal Surrey
Regiment.

RMAs Collection, Sandhurst
Tel: Camberley (0276) 63344 ext 489

Royal Army Ordnance Corps Museum,
Blackdown Barracks, Deepcut, Nr Camberley
Tel: Aldershot (0252) 24431 ext 5515
Uniforms, weapons and bomb disposal items
including the 'wheelbarrow'.

Staff College Museum, Sandhurst
Tel: Camberley (0276) 63344 ext 2602

Women's Royal Army Corps Museum,
WRAC Centre, Queen Elizabeth Park,
Guildford
Tel: Aldershot (0252) 24431 ext 8583
Women in the army since 1917.

TYNE AND WEAR

John George Joicey Museum, City Road,
Newcastle-upon-Tyne

Tel: 091-232 4562
Cavalry memorabilia from 15th/19th Hussars
and Northumberland Hussars.

Military Vehicle Museum, Exhibition Park
Pavilion, Newcastle-upon-Tyne
Tel: 091-281 7222
Mainly World War II vehicles.

WARWICKSHIRE

The Arms and Armour Museum, Poet's
Arbour, Sheep Street, Stratford-upon-Avon
Tel: Stratford-upon-Avon (0789) 293453
Weapons of every variety from crossbows and
early handguns to modern arms.

The Lunt Roman Fort, Coventry Road,
Baginton
Tel: Coventry (0203) 25555 ext 2315
Reconstructed Roman fort in original
dimensions; many Roman relics.

Midland Air Museum, Coventry Airport,
Baginton
Tel: Coventry (0203) 301033
Military aircraft of all kinds including the
most recent.

**Regimental Museum of the Queen's Own
Hussars,** The Lord Leycester Hospital, High
Street, Warwick
Tel: Warwick (0926) 492755
The Queen's Own Hussars were formed by
the amalgamation of the 3rd King's Own
Hussars and the 7th Queen's Own Hussars,
both of which have a distinguished history.

Warwickshire Yeomanry, The Court
House, Jury Street, Warwick
Tel: Warwick (0926) 492212
Yeomanry items.

WEST SUSSEX

Corps of Military Police Museum,
Roussillon Barracks, Chichester
Tel: Chichester (0243) 786311
400 years of army policing.

WEST YORKSHIRE

Bolling Hall Museum, Bowling Hall Road,
Bradford
Tel: Bradford (0274) 72345
Displays of the West Yorkshire Regiment

including the 'Bradford Pals' of the Somme battles.

City Museum, Municipal Buildings, Leeds
Tel: Leeds (0532) 462632
Leeds Rifles memorabilia.

The Duke of Wellington's Regiment, Bankfield Museum, Akroyd Park, Halifax
Tel: Halifax (0422) 54823
Extensive displays in period settings. Relics of the Duke of Wellington; material from associated regiments.

WILTSHIRE

Duke of Edinburgh's Royal Regiment (Berkshire and Wiltshire), 58 The Close, Salisbury
Tel: Salisbury (0722) 336222 ext 2683
Mainly Berkshire and Wiltshire, but also volunteer and militia items.

Museum and Art Gallery, Bath Road, Swindon
Tel: Swindon (0793) 26161 ext 3129
Wiltshire Yeomanry collection.

School of Infantry Museum, Warminster
Tel: Warminster (0985) 214000 ext 2487
2,000 weapons of every type from revolvers to anti-tank guns. Strictly by previous appointment only.

WALES

DYFED

Carmarthen Museum, Abergwili, Carmarthen
Tel: Carmarthen (0267) 231691
Mainly militia and volunteer items.

Castell Henllys Iron Age Fort, Texodus LDT, Pant Glas, Felindre Farchog, nr Crymych
Tel: (0239) 79319
This fort is being reconstructed on actual excavations.

Pembroke Yeomanry Trust, Castle Museum and Art Gallery, The Castle, Haverfordwest
Tel: Haverfordwest (0437) 3708
Exhibits of various yeomanry and militia regiments.

Tenby Museum, Castle Hill, Tenby
Tel: Tenby (0834) 2809
Section devoted to the Castlemartin Yeomanry.

GWENT

Caerleon Roman Fortress, Caerleon, Gwent
Tel: (0633) 422518
Once the Roman Fortress of Isca, headquarters of the 2nd Augustan Legion. Includes a new visitors' centre.

The Nelson Museum, Monmouth
Tel: Monmouth (0600) 713519
Principally items concerning Admiral Lord Nelson.

Royal Monmouthshire Royal Engineers, The Castle, Monmouth
Tel: Monmouth (0600) 2395
Regimental memorabilia.

GWYNEDD

Plas Newydd (National Trust), Llanfairpwll, Anglesey
Tel: (0248) 714798
This late eighteenth-century house contains relics of the Battle of Waterloo.

Regimental Museum of the Royal Welch Fusiliers, The Queen's Tower, Caernarfon Castle, Caernarfon
Tel: Caernarfon (0286) 233821
Includes a Russian 16-pdr gun captured at the Battle of the Alma in 1854. The spelling of 'Welch' was 'Welsh' from 1881 to 1920, but had been 'Welch' earlier.

Segontium Roman Fort Museum, Caernarfon
Tel: (0286) 5625
A purpose built gallery on the site of the Roman Fort excavated by Sir Mortimer Wheeler in the 1920s. Many notable artefacts.

POWYS

Powis Castle, Welshpool
Tel: (0938) 4336
A majestic thirteenth-century castle,
containing a fine collection of tapestries,
furniture and pictures.

**Regimental Museum of the South Wales
Borderers and Monmouthshire
Regiment**, The Barracks, Brecon
Tel: Brecon (0874) 3111 ext 310
The South Wales Borderers is now
amalgamated with the Welch Regiment which
has a separate museum at Cardiff. Items from
the Zulu War, audio-visual display.

SOUTH GLAMORGAN

Historic Aircraft Collection, RAF St Athan,
Barry
Tel: Barry (0446) 798798 ext 4878
Historic aircraft and engines. Strictly by
appointment only.

The Welch Regiment Museum, The Black
and Barbican Towers, Cardiff Castle, Cardiff
Tel: Cardiff (0222) 29367
Many items from the regiment which began in
1719 as the 41st (Royal Invalids) Regiment
and is now, with the SWB, the Royal Regiment
of Wales.

SCOTLAND

BORDERS

Coldstream Museum, Coldstream,
Berwickshire
Tel: (0890) 2630
Small local history museum. Includes material
of Coldstream Guards.

CENTRAL

**The Argyll and Sutherland Highlanders
Regimental Museum**, Stirling Castle,
Stirling FK8 1EH
Tel: Stirling (0786) 75165
Relics and paintings of the Argylls (91st and
93rd) including 'The Thin Red Line' in the
Crimea.

GRAMPIAN

**Gordon Highlanders Regimental
Museum**, Regimental Headquarters,
Viewfield Road, Aberdeen
Tel: Aberdeen (0224) 318174
Memorabilia of various campaigns and battles
and a Victoria Cross exhibition.

HIGHLAND

Clan Cameron Museum, Achnacarry, Spean
Bridge, Inverness-shire
Tel: (0397) 772 473
A reconstructed 17th century croft house.
Sections on the Queen's Own Cameron
Highlanders and the Commandos, who trained
at Achnacarry during World War II.

Culloden Moor and Visitor Centre, By
Inverness
Tel (Visitor Centre): (0463) 7900607
Site of the Battle of Culloden (16 April 1746).
Features of interest include the Graves of the
Clans, communal burial places with simple
headstones bearing individual clan names: the
Well of the Dead, a single stone with the
inscription 'The English were buried here';
Old Leanach Farmhouse, now restored as a
battle museum; and the huge Cumberland
Stone, from which the victorious Duke of
Cumberland is said to have viewed the scene.

**Queen's Own Highlanders Regimental
Museum**, Fort George, by Ardersier,
Inverness-shire
Tel: (0463) 224380
Items from the Seaforth Highlanders and the
Cameron Highlanders (now amalgamated as
the Queen's Own Highlanders) and also the
Lovat Scouts.

West Highland Museum, Cameron Square,
Fort William
Tel: Fort William (0397) 2169
Items from the Cameron Highlanders and
others.

LOTHIAN

Huntly House Museum, Canongate, Royal
Mile, Edinburgh
Tel: 031-225 2424 ext 6689
Items from the local militia and personal
relics of Field Marshal Earl Haig.

Royal Museum of Scotland, Queen Street, Edinburgh
Tel: 031-225 7534
Prehistoric, Roman and traditional Scottish weapons.

Penicuik Scottish Infantry Depot Museum, Glencorse Barracks, Milton Bridge, Penicuik
Tel: Penicuik (0968) 72651 ext 23
Weapons of every variety.

The Royal Scots Dragoon Guards Display Room, The Castle, Edinburgh
Tel: 031-336 1761
Display of pictures, badges, brassware and other historical relics of the regiment.

Royal Scots Museum, The Castle, Edinburgh
Tel: 031-336 1761 ext 4625
Regimental relics of the Royal Scots, the oldest regular regiment in the British Army (infantry).

Scottish United Services Museum, Crown Square, The Castle, Edinburgh
Tel: 031-225 7534
Large, impressive museum with well arranged displays of every aspect of Scottish military history.

STRATHCLYDE

The Burrell Collection, Pollock Country Park, 2060 Pollockshaws Road, Glasgow
Tel: 041-649 7151
Arms and armour.

The Cameronians (Scottish Rifles) Regimental Museum, Mote Hill, Muir Street, Hamilton,
Tel: Hamilton (0698) 428688 ~~283981~~ 328252
Weapons, uniforms, medals, banners and documents relating to the regiment.

Combined Operations Museum, Cherry Park, Inveraray Castle, Inveraray
Tel: (0499) 2205
Commando training took place in this rugged area and the story of the development of Combined Operations, including the resulting raids, is on display here.

Culzean Castle and Country Park, Maybole, Ayrshire
Tel: Kirkoswald (065 56) 274/269
Elaborate displays of weapons supplied to the West Lowland Regiment during the Napoleonic Wars.

Dean Castle, Dean Road, Kilmarnock
Collection of European arms and armour.

Glasgow Art Gallery and Museum, Kelvingrove, Glasgow
Tel: 041-357 3929
Arms and armour

The Hunterian Museum, Glasgow University, Glasgow
Tel: 041-330 4221
Mainly Roman army relics.

Royal Highland Fusiliers Regimental Museum, 518 Sauchiehall Street, Glasgow
Tel: 041-332 0961
Items from regiment formed by the amalgamation of the Royal Scots Fusiliers and the Highland Light Infantry, two renowned regiments.

The Rozelle Galleries, Rozelle Park, Monument Road, Alloway by Ayr
Tel: Ayr (0292) 45447/43708
Yeomanry items.

TAYSIDE

The Black Watch Museum, Balhousie Castle, Hay Street, Perth
Tel: Perth (0738) 21281 ext 8530
Memorabilia of the Black Watch.

Blair Castle, Blair Atholl, Pitlochry, Perthshire
Tel: Blair Atholl (079 681) 207
Items of the Atholl Highlanders, a private army of the duke.

The Scottish Horse Museum, The Cross, Dunkeld, Perthshire
Tel: Colonel Lisle (0350) 9934205
Items from the Scottish Horse Regiment.

NORTHERN IRELAND

CO ANTRIM

Home Front Heritage Centre, 9 Waring
Street, Belfast BT1 2DW
Tel: Belfast (0232) 320392
Exhibits from World War II include tin hats,
gas masks, landgirl's uniform and an
incendiary bomb dropped on Belfast in 1941.

The Royal Ulster Rifles, RHQ The Royal
Irish Rangers, 5 Waring Street, Belfast
Tel: Belfast (0232) 232086
Memorabilia of the Royal Ulster Rifles.

Royal Ulster Rifles' Regimental Museum,
War Memorial Building, 5 Waring St, Belfast
Tel: Belfast (0232) 232086
Preserved are the uniforms, badges, medals,
weapons and records of the 83rd and 86th
Foot Regiments, the Royal Irish Rifles and the
Royal Ulster Rifles – from 1793–1968. Four of
the Victoria Crosses on display were awarded
for heroism during the Indian Mutiny and
three during World War I. Battle of the
Somme records and memorabilia.

Ulster Museum, Botanic Gardens, Belfast
Tel: Belfast (0232) 381251
Located on first floor. Small collection of
memorabilia of Irish regiments. Uniforms date
from 1780 (worn by the Irish Volunteers) right
up to one worn at the Falkland's Campaign
(1982), and among the weapons is an 1844
pikehead. A special exhibit is the regimental
flag of the Belfast Yeomanry (1803).

CO ARMAGH

Armagh County Museum, The Mall East,
Armagh, Co Armagh
Tel: Armagh (0861) 523070
Small collection of memorabilia of The
Armagh Volunteers (1792), The Yeomanry
(1796), The Militia (19th-c), North Irish Horse
and Royal Irish Fusiliers (both 20th-c).

Royal Irish Fusiliers Regimental Museum,
Sovereign's House, The Mall, Armagh
Tel: Armagh (0861) 522911
Founded in 1793, this regiment was
amalgamated with the Royal Inniskilling
Fusiliers and the Royal Ulster Rifles in 1968
to become the 3rd Battalion Royal Irish
Rangers. Many manuscripts and war diaries.

CO FERMANAGH

Battle of the Atlantic Exhibition, Castle
Archdale Country Park, Irvinestown
BT94 1PP
Tel: Irvinestown (036 56) 21588.
In an outhouse in the courtyard of the park,
there is a World War II Exhibition featuring
the Battle of the Atlantic. During the battle
the RAF flew Sunderland and Catalina
seaplanes from here and these are
prominently featured.

**Regimental Museum of the Royal
Inniskilling Fusiliers**, The Castle,
Enniskillen
Tel: Enniskillen (0365) 23142
History of the regiment from its formation in
1689 to its amalgamation with the Royal Irish
Rangers in 1968.

CHANNEL ISLANDS

JERSEY

Elizabeth Castle, St Aubin's Bay, Nr St
Helier
Tel: Jersey (0534) 23971
British and local militia items.

German Underground Hospital, St
Lawrence
Tel: Jersey (0534) 63442
Horrific memorial to the German occupation
of the Channel Islands for it was dug out by
Russian civilians used as slave lavour and
took 2½ years. German relics and video
displays.

La Hougue Bie Museum, Grouville
Tel: Jersey (0534) 53823
Complete German bunker containing relics of
the German occupation.

Mont-Orgueil Museum, Gorey, Nr St Helier
Tel: Jersey (0534) 53292
History of Mont-Orgueil Castle, strategically
situated with view of entire Jersey coastline.

Museum of Nazi Equipment, St Peter's
Bunker
Tel: Jersey (0534) 33825
Examples of the equipment the British soldier
had to face in defeating Nazi Germany.

INDEX

Abyssinia, 122
Aden, 173
Afghanistan, 93
Alamein, Battle of, 171
Allenby, General, 137
Alma, Battle of, 99
America, North, 54
Arnhem, 159
Arras, 141
Artillery, 35
Ashanti Wars, 124
Assegais, 110
Auchinleck, 152

Bailey bridges, 166
Ball, Captain Albert, VC, 145
Bands, 62
Barracks, 61
Bayonets, 40
Boer War, (First), 113
Boots, 121
Boxer War, 121-2
Bren Guns, 125
Bromhead, Lt, VC, 112
Bronze Age, 17
'Brown Bess', 125
Burma War, 92, 93

Canada, 64
Cat-o'-nine-tails, 44
Celts, 18
Chard, Lt, VC, 112
Charge of the Light and Heavy
    Brigades, 100-5
Chavasse, Captain N. G., VC, 133
Chindits, 163
Churchill, 150
Clive, 64
Colours, 62
Commandos, 113, 150
Commissions, 46
Cooking, 59
Crimean War, 98-108
Culloden, 53

Dervishes, 119-20
Diamonds, 109
Diehards, 77
Drink, 60
Dunkirk, 149

East India Company, 55
Edgehill, 32
'Enigma', 169

Falklands War, 176
Flintlocks, 29
Food, 57
French Revolution, 72

Gallipoli, 136
Gas, 134
Gordon, General Charles, 119
Guards Armoured Division, 15
Gulf War, 176
Gunpowder, 36
Guns, 125
Gurkhas, 89

Haig, Field Marshal Earl, 132
Harris, Rifleman, 48
Hastings, Battle of, 24
Home Guard, 151
Humour, 13

Independence, War of American,
    68-9
Indian Mutiny, 97
Inkerman, Battle of, 97
Iron Age, 17
Isandhlwana, Battle of, 110

Japanese Army, 159-63
Jenkin's Ear, War of, 50-51

Kabul, 95
Kenilworth, 36
Khartoum, 119
Killiecrankie, 39
Kipling, 67
Kitchener, 117
Kohima, Battle of, 162
Korean War, 172
Kruger, 115
Kumasi, 124

Ladysmith, Siege of, 116
Le Haye Sainte, 81
Lee-Enfield Rifles, 125
Longbow, 9
Long Range Desert Group (LRDG),
    156

Machine-guns, 128
'Mad Mullah', 123
Marlborough, 40-1
Matchlock, 29
Mayne, Lt Colonel, 156
Medical Miracles, 10
Mercenaries, 25
Mesopotamioa, 136

Napoleon, 72-3
Napoleon, Prince Louis, 113
Naseby, Battle of, 33
Nelson, Admiral Lord, 72-3
New Model Army, 26-34
Nightingale, Florence, 107-8
Normans, 24

Omdurman, Battle of, 121

Panzers, 152
Passchendaele, 129
Pay, Soldiers', 59
'Phantom', 168
Picton, General, 81, 86
'Purple', 169

Quatre Bras, 83

R.A.F., 151
Raglan, Lord, 99-105
Redcoats, 42
Rifle Brigade, 70
Roberts, General 'Bobs', 96
Romans, 19-20
Rommel, Erwin, 152
Rorke's Drift, Battle of, 111-12
Royal Hospital, 42, 45

Salamanca, Battle of, 78
Sandhurst, 57, 113
S.A.S., 156, 173
Self-Loading Rifle (SLR), 127
Slave Trade, 123
Somme, The, 137-41
South Africa, 109-20
Special Operations Executive (SOE),
    151
Stirling, David, 156

Tanks, First use of, 140
Thin Red Line, 100
Torres Vedras, Lines of, 75
Trafalgar, Battle of, 73

'Ultra', 169
Ulundi, Battle of, 113
United States, see America, North

Verdun, 139
Vickers Machine Gun, 128
Victoria Cross, 133
Vikings, 21
Vittoria, Battle of, 79

Washington, George, 68
Waterloo, 81
Wives, 61
Wolfe, General James, 65
Woolwich, 57

Yemen, 173
Ypres, Battle of, 129
'Y' Service, 168

Zulu War, 110-13